中高职衔接贯通培养计算机类系列教材

Java面向对象程序设计

何　鑫　杨翠萍　主　编

孙守梅　张业男　副主编

薛永三　主　审

化学工业出版社

·北京·

本书作为中高职衔接贯通培养计算机类系列教材，是Java语言类课程第二阶段的深入教学用书，分为10个项目，项目1类与对象、类的方法、类的封装；项目2继承和多态；项目3异常处理；项目4集合与泛型；项目5 Java数据库连接；项目6 Java图形用户界面；项目7 Java中的输入/输出流技术；项目8多线程机制；项目9网络编程；项目10综合实战项目。

本书内容安排合理，逻辑性强，讲解循序渐进，通俗易懂，符合三二分段的中高职两个学习阶段的学生认知过程及学习规律，适合高等职业院校计算机及信息工程或相关专业教材或参考书，也可供相关领域的读者参考。

图书在版编目（CIP）数据

Java面向对象程序设计 / 何鑫，杨翠萍主编. —北京：化学工业出版社，2017.2

中高职衔接贯通培养计算机类系列教材

ISBN 978-7-122-28881-3

Ⅰ. ①J… Ⅱ. ①何… ②杨… Ⅲ. ①Java语言-程序设计-职业教育-教材 Ⅳ. ①TP312.8

中国版本图书馆CIP数据核字（2017）第008682号

责任编辑：张绪瑞 廉 静　　装帧设计：刘丽华

责任校对：宋 玮

出版发行：化学工业出版社（北京市东城区青年湖南街13号 邮政编码100011）

印　　装：三河市延风印装有限公司

787mm×1092mm 1/16 印张15¾ 字数384千字 2017年3月北京第1版第1次印刷

购书咨询：010-64518888(传真：010-64519686) 售后服务：010-64518899

网　　址：http://www.cip.com.cn

凡购买本书，如有缺损质量问题，本社销售中心负责调换。

定　　价：38.00元

中高职衔接贯通培养计算机类系列教材
编审委员会

编写说明

黑龙江农业经济职业学院2013年被黑龙江省教育厅确立为黑龙江省首批中高职衔接贯通培养试点院校，在作物生产技术、农业经济管理、畜牧兽医、水利工程、会计电算化、计算机应用技术6个专业开展贯通培养试点，按照《黑龙江省中高职衔接贯通培养试点方案》要求，以学院牵头成立的黑龙江省现代农业职业教育集团为载体，与集团内20多所中职学校合作，采取“二三分段”（两年中职学习、三年高职学习）和“三二分段”（三年中职学习、两年高职学习）培养方式，以“统一方案（人才培养方案、工作方案）、统一标准（课程标准、技能考核标准），共享资源、联合培养”为原则，携手中高职院校和相关行业企业协会，发挥多方协作育人的优势，共同做好贯通培养试点工作。

学院高度重视贯通培养试点工作，紧紧围绕黑龙江省产业结构调整及经济发展方式转变对高素质技术技能人才的需要，坚持以人的可持续发展需要和综合职业能力培养为主线，以职业成长为导向，科学设计一体化人才培养方案，明确中职和高职两个阶段的培养规格，按职业能力和素养形成要求进行课程重组，整体设计、统筹安排、分阶段实施，联手行业企业共同探索技术技能人才的系统培养。

在贯通教材开发方面，学院成立了中高职衔接贯通培养教材编审委员会，依据《教育部关于推进中等和高等职业教育协调发展的指导意见（教职成[2011]9号)》及《教育部关于“十二五”职业教育教材建设的若干意见（教职成[2012]9号)》文件精神，以“五个对接”（专业与产业对接、课程内容与职业标准对接、教学过程与生产过程对接、学历证书与职业资格证书对接、职业教育与终身学习对接）为原则，围绕中等和高等职业教育接续专业的人才培养目标，系统设计、统筹规划课程开发，明确各自的教学重点，推进专业课程体系的有机衔接，统筹开发中高职教材，强化教材的沟通与衔接，实现教学重点、课程内容、能力结构以及评价标准的有机衔接和贯通，力求“彰显职业特质、彰显贯通特色、彰显专业特点、彰显课程特性”，编写出版了一批反映产业技术升级、符合职业教育规律和技能型人才成长规律的中高职贯通特色教材。

系列贯通教材开发体现了以下特点：

一是创新教材开发机制，校企行联合编写。联合试点中职学校和行业企业，按课程门类组建课程开发与建设团队，在课程相关职业岗位调研基础上，同步开发中高职段紧密关联课程，采取双主编制，教材出版由学院中高职衔接贯通培养教材编审委员会统筹管理。

二是创新教材编写内容，融入行业职业标准。围绕专业人才培养目标和规格，有效融入相关行业标准、职业标准和典型企业技术规范，同时注重吸收行业发展的新知识、新技术、新工艺、新方法，以实现教学内容的及时更新。

三是适应系统培养要求，突出前后贯通有机衔接。在确定好人才培养规格定位的基础上，合理确立课程内容体系。既要避免内容重复，又要避免中高职教材脱节、断层问题，要着力突出体现中高职段紧密关联课程的知识点和技能点的有序衔接。

四是对接岗位典型工作任务，创新教材内容体系。按照教学做一体化的思路来开发教材。科学构建教材体系，突出职业能力培养，以典型工作任务和生产项目为载体，以工作过程系统化为一条明线，以基础知识成系统和实践动手能力成系统为两条暗线，系统化构建教材体系，并充分体现基础知识培养和实践动手能力培养的有机融合。

五是以自主学习为导向，创新教材编写组织形式。按照任务布置、知识要点、操作训练、知识拓展、任务实施等环节设计编写体例，融入典型项目、典型案例等内容，突出学生自主学习能力的培养。

贯通培养系列教材的编写凝聚了贯通试点专业骨干教师的心血，得到了行业企业专家的支持，特此深表谢意！作为创新性的教材，编写过程中难免有不完善之处，期待广大教材使用者提出批评指正，我们将持续改进。

中高职衔接贯通培养计算机类系列教材编审委员会

2016 年 6 月

FOREWORD 前言

Java是由Sun Microsystems公司于1995年推出的可以编写跨平台应用软件的面向对象的高级程序设计语言。2010年Sun Microsystems公司被Oracle公司收购。现今Java是几乎所有类型的网络应用程序的基础，也是开发和提供嵌入式和移动应用程序、游戏、基于Web的内容与企业软件的全球标准。Java语言在全球有超过900万的开发人员，能够高效地开发、部署功能强大的应用程序和服务。鉴于其在软件开发方面的霸主地位，各高校已将其作为计算机类学生必修的课程之一。

本教材注重培养学生综合职业能力，教材注重由浅入深、由点到面，以能力为主线的整体设计思路，重新组合课程，节约之前重复学习的时间，做到知识结构的连贯性，并遵循从学生专业能力、方法能力、社会能力和发展能力角度出发，内容循序渐进、深入浅出，精心设计每一个示例，结构安排更为合理，使读者准确把握Java的知识点。本教材在讲解一项任务时按照“需求分析”、“技能解析”、“知识解析”、“编码实施”、“调试运行”和“维护升级”的职业情境为主线，紧跟相关的实例演示，使学习者综合应用已经学过的主要知识，以期达到培养出企业真正急需人才的目的。

本书作为中高职衔接贯通培养教材Java语言类课程第二阶段的深入教学，分为10个项目，以下是每一项目的简单介绍。

项目1：类与对象、类的方法、类的封装；让学生理解面向对象程序设计思想，能从结构化程序设计思路转变到面向对象程序设计思想上，介绍类的定义和对象的使用；能够在编写程序时可以熟练地使用方法，并且能够对程序进行正确的封装。

项目2：继承和多态；介绍this、super和final关键字的使用场景和作用。可以利用接口做更深层次的抽象。

项目3：异常处理；学会异常捕获、异常处理、抛出异常的方法，能够利用异常处理机制处理程序中可能出现的异常。

项目4：集合与泛型；重点介绍ArrayList、HashMap两种集合的使用场景，学会使用迭代器遍历集合，介绍泛型的基本使用。

项目5：Java数据库连接；介绍JDBC编程的基本步骤，利用JDBC技术实现对不同类型数据库(access ，SQL Server, MySql)的操作，掌握JDBC编程在实际项目中的应用，为Java数据库系统开发打下良好的基础。

项目6：Java图形用户界面；介绍了容器、布局管理器、常用组件和Java的事件处理机制，对于比较复杂的组件都给出了很实用的例子。

项目7：Java中的输入/输出流技术。

项目8：多线程机制；首先介绍Java线程的运行机制，然后介绍多线程的基本概念与创建、启动方法，以及如何对多个线程进行调度、同步和通信的基本知识。

项目9：网络编程；介绍网络编程中的基本概念，理解并比较TCP协议与UDP协议两种网络编程的实现方式；能分别使用Socket类与ServerSocket类来创建客户端程序与服务端程序。

项目10：综合实战项目，巩固和提升学生对所学知识的综合应用能力。主要介绍学生信息管理系统的实现过程，如系统的需求分析、概要设计、数据库设计、模块实现和系统测试等。最后编码实现该项目。

本书内容安排合理，逻辑性强，讲解循序渐进，通俗易懂，符合三二分段的中高职两个学习阶段的学生认知过程及学习规律，适合高等职业院校计算机及信息工程或相关专业教材或参考书，也可供相关领域的读者参考。

本教材由黑龙江农业经济职业学院何鑫、黑龙江省依兰县职业中学杨翠萍担任主编，负责制定编写大纲和全书统稿工作，黑龙江农业经济职业学院孙守梅、张业男担任副主编。具体分工为：项目 1、项目 2 中的任务 1、任务 2 由黑龙江农业经济职业学院张业男负责编写；项目 3 由黑龙江农业经济职业学院于瀛军负责编写；项目 4 中的任务 1 由哈尔滨学院刘磊负责编写；项目 4 中的任务 2 及项目实训与练习由依兰县职业中学杨翠萍负责编写；项目 5、项目 6 由黑龙江农业经济职业学院孙守梅负责编写；项目 2 中的任务 3 及项目实训与练习、项目 7、项目 8、项目 9、项目 10 由黑龙江农业经济职业学院何鑫负责编写。编写团队在这一年多的编写过程中付出了很多辛勤的汗水，尽管我们尽了最大的努力，但教材中难免会有不妥之处，欢迎各界专家和读者朋友们提供宝贵意见和建议，我们不胜感激！

编　者

CONTENTS 目录

项目 7 输入输出流

项目 8 Java 的分身术：多线程机制

欢迎来到另一个世界

项目目标

本项目主要内容是类与对象、类的方法、类的封装；通过本任务的学习让学生理解面向对象程序设计思想，能从结构化程序设计思路转变到面向对象程序设计思想上，掌握类的定义和对象的使用；能够在编写程序时熟练地使用方法，并且能够对程序进行正确的封装。

项目内容

对日常生活中常见的汽车进行抽象，得到汽车类；利用汽车类创建不同的汽车类对象。

设计一个狗类，利用方法限定狗的属性 size，属性不能小于 0，当 size 大于 15 和小于 15 时有不同的叫声。

任务1　了解另一个世界：类与对象

这是一个新奇的世界，在这里所有的物质都是对象。在前面的课程中大多数时候，我们的程序代码都是写在 main( )方法中的；这不是这个世界的做法。我们将离开“过程化”的世界，进入“面向对象”的世界。感受“面向对象”开发的乐趣吧。

首先通过下面的趣事了解“面向对象”世界的人是如何设计程序的。

从前有一个软件开发工作室，有两个程序员被指派去开发一个程序，为了让这两个程序员工作更有激情，项目经理要求他们比赛，赢的人可以将自己的电脑升级成 8GB 内存，将可以极大提高工作效率。这两名程序员分别是“过程化”程序开发高手——小强，以及“面向对象”程序开发高手——小明。两个人的战争一触即发。

✧ 需求分析

画出圆三角形（Triangle）、矩形（Rectangle）和五边形（Pentagon）。当用户点选图形时，图形需要顺时针旋转 180°，并放大三倍。如图 1-1 所示。

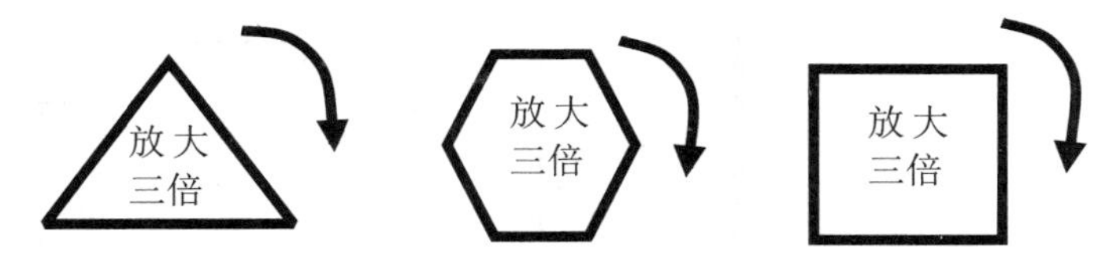

图 1-1　程序设计要求

“过程化”高手小强的分析：根据需求我需要 rotate（旋转）和 enlarge（放大）两个动作，也就是需要用到两个方法。

“面向对象”高手小明的分析：五边形、四边形和三角形都是图形。这些图形都有自己的动作，也就是都有自己的方法。

下面是他们编写的代码，为方便理解，具体的语句用注释代替。

小强：非常快速地写出了两个方法。

```
void rotate(){
    //旋转180°
}
void enlarge(){
    //放大三倍
}
```

小明：根据三个不同的图形写出了三个类。

Triangle
(三角形类)

```
void ro
//旋转180
}
void enl
//放大三
}
```

Rectangle
(四边形类)

```
void rotate(
//旋转180°
}
void enlarge
//放大三倍
}
```

Pentagon
(五边形类)

```
void rotate(){
//旋转180°
}
void enlarge(){
//放大三倍
}
```

“过程化”高手小强和“面向对象”高手小明这一轮比拼中，小强的开发速度似乎更快一些。看起来小明的开发方式看起来更复杂一些。

这时项目经理修改了需求，加入圆形，如图 1-2，两个人又开始修改。

图 1-2　圆形设计要求

小强修改后的代码：

```
void rotate(){
    //旋转180°
}
void enlarge(){
    //如果是圆形
        //放大两倍
    //否则
        //放大三倍
}
```

小明修改后的代码：

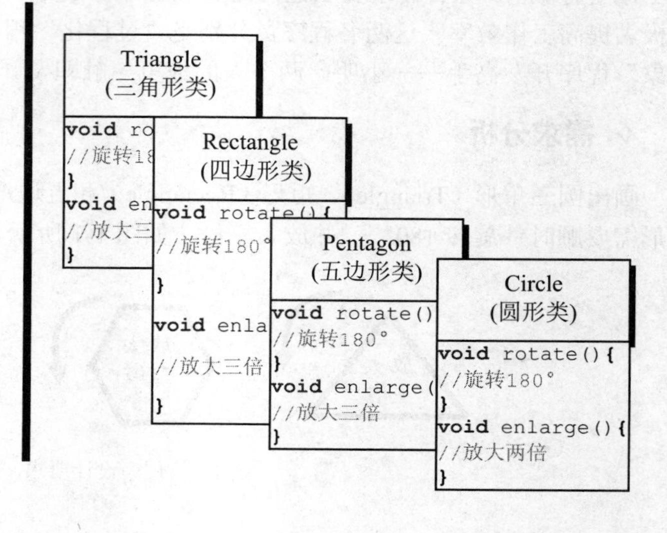

这时会发现，两个人的速度差不多了。要知道在这个程序开发过程中修改需求是经常发生的，且不可避免。紧接着项目经理又提出新的修改要求，由于圆形在平面上旋转效果不明显，所以需要让圆形绕轴旋转，如图1-3所示。

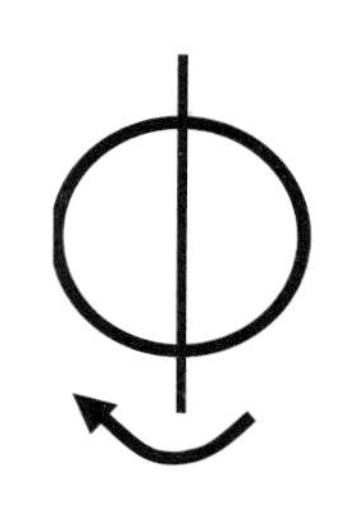

图1-3　圆形设计要求

这时“过程化”高手小强在修改程序时就比较吃力了，因为他的程序是一个整体，每一次修改都是对整个程序的修改，随着程序的功能越来越多，需要修改和考虑的地方也随之增加，也就越来越吃力。而“面向对象”高手小明就要好很多，因为他只需要修改对应的“类”就可以了。下面是他们的修改结果：

小强修改后的代码：

```
void rotate(){
//如果是圆形
        //绕轴旋转
    //否则
        //旋转180°
}
void enlarge(){
    //如果是圆形
        //放大两倍
//否则
        //放大三倍

}
```

小明只需要修改一个“类”：

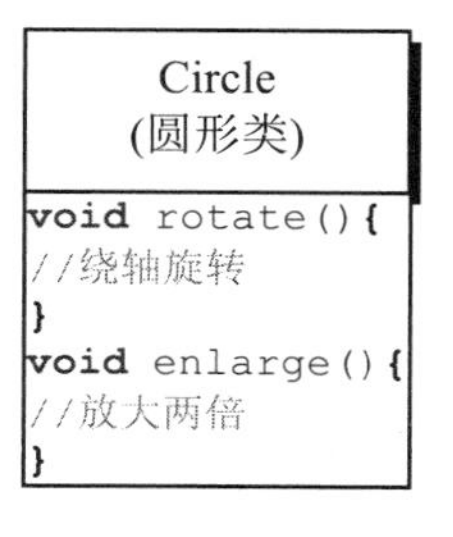

暂时来看程序的修改不大，无论“过程化”还是“面向对象”，作为编程高手，小强和小明知道需求的修改是家常便饭，随着程序逐渐庞大，走“面向对象”路线的小明优势会越来越明显。所以最终“面向对象”高手小明会胜出。

✧ 知识准备

1．技能解析

这种“面向对象”的程序该如何设计呢？

首先需要用“面向对象”的角度去看待问题，将所有的事物都看做“对象”。提取它们的特征。

提取的特征通常包括：

（1）已知的属性。

（2）执行的动作。

如图1-4所示，分析这个按钮有哪些特征能够提取。

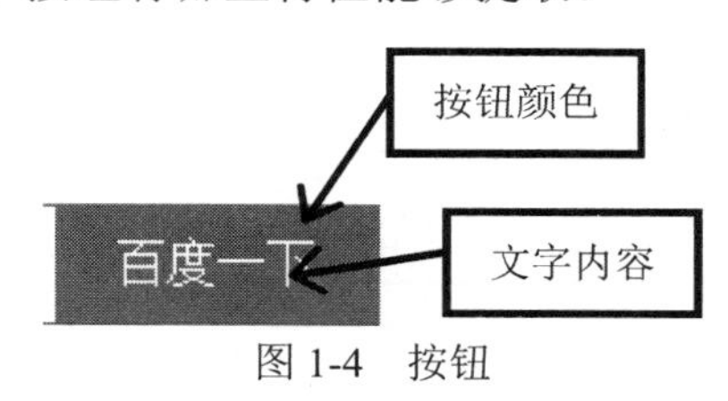

图1-4　按钮

■ 已知的属性：按钮的颜色（color）和文字的内容（label）。

■ 执行的动作：点击（click）、改变按钮颜色、改变文字内容等。

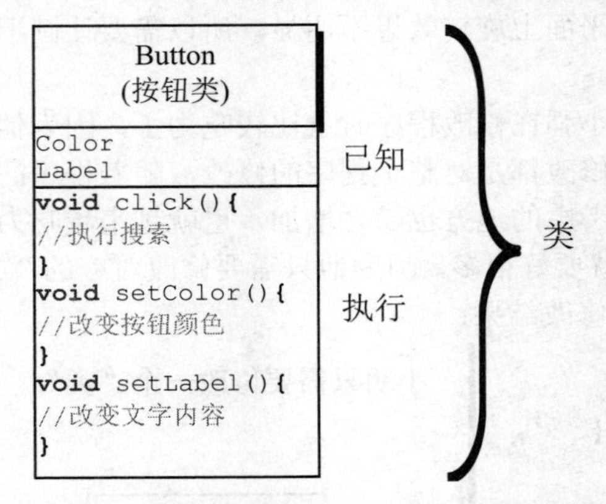

这个提取特征的过程在“面向对象”世界中叫**抽象**。抽象之后形成的叫做**类**。已知的属性称为**属性**，执行的动作称为**方法**。

在图 1-5 中填入汽车的属性和方法（各填两个）：

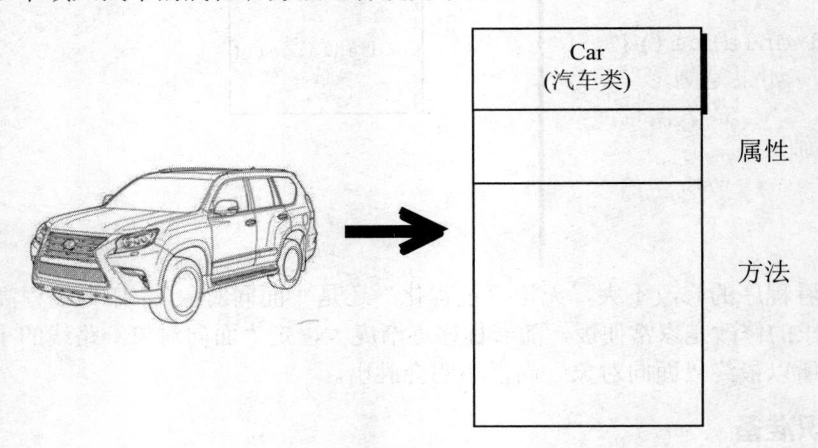

图 1-5　汽车的属性和方法

当你填入相关内容的时候也就完成了一次抽象，你成功地抽象出了一个汽车类。其实你抽象出的汽车类，只要给属性不同的值，就可以成为不同的汽车。比如下面抽象出的汽车类。

想象一下符合图 1-6 所有要求的车是什么样的？

```
Car
(汽车类)
Type   //型号
Color  //颜色
start()  //发动
brake()  //刹车
```

图 1-6　汽车类图

只需要它们的属性不同的值，就能够表示不同的汽车。

这些不同的车称之为汽车类的**对象**（见图 1-7）。这也完成了一次从抽象到具体的过程。这些对象想要使用，通常需要两个类进行配合，一个是将要被操作的类（比如 Car 汽车类，Button 按钮类等），另一个是测试类。测试类中需要包含 main()方法。

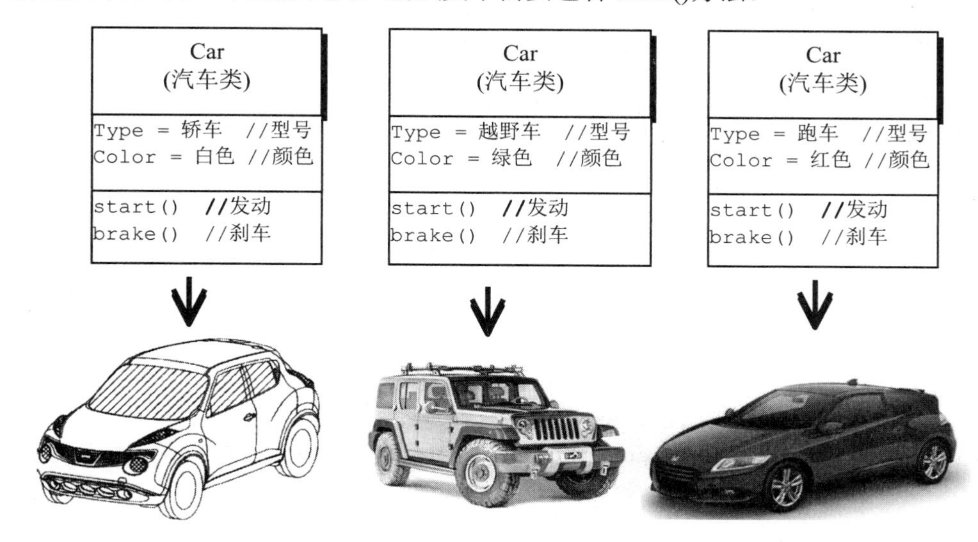

图 1-7 汽车对象

2．知识解析

类与对象是面向对象程序设计语言中的核心概念。类不是对象，类是对象的设计图。根据某个类创建出的对象都会有自己的属性和方法，虽然这些对象都是根据一个类创建出来的，但是这些对象的属性方法可以完全不同。比如，利用汽车类可以创建出许多型号、颜色、品牌等不同的汽车，你也可以利用按钮类创建出颜色、文字、大小不同的许多按钮，如图 1-8 所示。

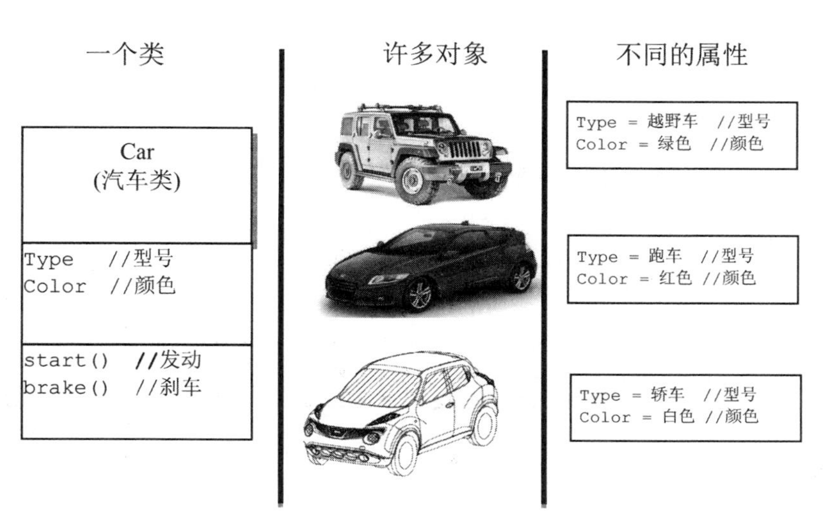

图 1-8 类与对象的关系

创建的对象的过程也被称为**实例化对象**。对于一个货真价实的面向对象程序来说，main()方法有两个用途：

（1）测试真正的类。

（2）启动 java 程序。

✧ 编码实施

将图 1-8 中三台车创建出来可以用这样的写法。

1．编写类

```
public class Car {
        String type; //车型
        String color;    //颜色

        public void start(){
             System.out.println(color+type+"汽车发动");
        }
        public void brake(){
             System.out.println(color+type+"正在踩刹车");
        }
}
```

2．在测试类中创建对象

```
public class Test {
     public static void main(String[] args) {
              //Car 测试代码
              Car car1 = new Car( );//创建轿车对象
              car1.color="白色";//给 color 属性赋值
              car1.type="轿车";//给 type 属性赋值

              car1.start();//执行汽车发动方法
              car1.brake();//执行踩刹车方法
     }
}
```

控制台输出如下图示例：

```
<terminated> Test [Java Application] D:\Program Files\Java
白色轿车汽车发动
白色轿车正在踩刹车
```

➢ 练一练：

仿照上面的过程，创建其他两个汽车类对象，并测试运行。

✧ 调试运行

1．将程序代码进行如下修改，在代码编写过程中不用 new 关键字创建对象，会出现如下所示的错误：

```
public class Test {
    public static void main(String[] args) {
        //Car测试代码
        car1.color="白色";//给color属性赋值
        car1.type="轿车";//给type属性赋值

        car1.start();//执行汽车发动方法
        car1.brake();//执行踩刹车方法
    }
}
```

car1 cannot be resolved
7 quick fixes available:
- Create local variable 'car1'
- Create field 'car1'
- Create parameter 'car1'
- Create class 'car1'
- Create constant 'car1'
- Change to 'Car' (chap01)
- Fix project setup...

我们得出结论，对象的使用跟变量类似，必须先创建再使用。

2．将程序进行如下修改，将 Car 类中的 brake()方法注释去掉。

```
public class Car {
    String type; //车型
    String color;   //颜色

    public void start(){
        System.out.println(color+type+"汽车发动");
    }
    /*
    public void brake(){
        System.out.println(color+type+"正在踩刹车");
    }
    */
}
```

此时在对象使用过程中会出现如下错误：

```
public class Test {
    public static void main(String[] args) {
        //Car测试代码
        Car car1 = new Car();//创建轿车对象
        car1.color="白色";//给color属性赋值
        car1.type="轿车";//给type属性赋值

        car1.start();//执行汽车发动方法
        car1.brake();//执行踩刹车方法
    }
}
```

The method brake() is undefined for the type Car
2 quick fixes available:
- Create method 'brake()' in type 'Car'
- Add cast to 'car1'

我们可以看出在类的编写过程中没有方法也是不能使用的。

✧ 维护升级

完善上述程序，创建三台车的对象。

```
public class Test2 {
    public static void main(String[] args) {
        //Car 测试代码
        Car car1 = new Car();//创建轿车对象
```

```
        Car car2 = new Car();//创建越野车对象
        Car car3 = new Car();//创建跑车对象

        car1.color="白色";//给 color 属性赋值
        car1.type="轿车";//给 type 属性赋值
        car1.start();//执行汽车发动方法
        car1.brake();//执行踩刹车方法

        //操作越野车对象
        car2.color="绿色";
        car2.type="越野车";
        car2.start();
        car2.brake();

        //操作跑车对象
        car3.color="红色";
        car3.type="跑车";
        car3.start();
        car3.brake();
    }
}
```

控制台输出信息：

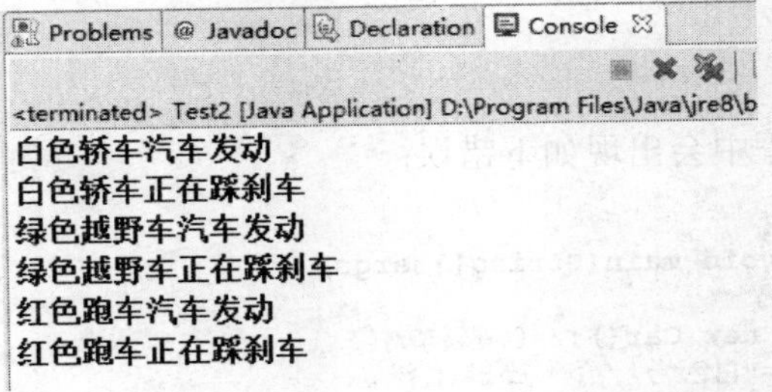

从上面的程序中能够看到给三个对象的属性赋予不同的值，它们的输出结果也不同。对象和对象之间是互不影响的，相互独立，即使它们都是由一个类创建出来的。

所有的对象都是由 new 关键字创建出来的，它们都属于**引用类型**。一个对象如果想要被使用，那么必须要有下图中的实例化过程。

```
Car car1 = new Car();
```

①　③　②

1．声明引用类型变量

分配空间，存储引用类型变量，并且它的名字为 car1，它的类型被永久固定为 Car。此时仅仅有了一个名字，也就是说这个引用变量还无法使用任何的属性或方法。

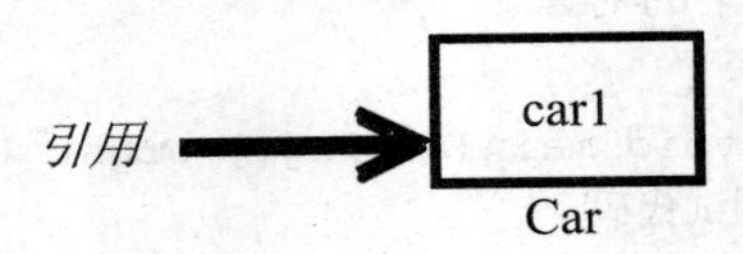

2．创建对象

分配空间存储新建立的 Car 对象。此时有了已经有了的对象，属性和方法也准备好了，不过引用类型变量 car1 和新建对象之间还没有什么联系，这些属性和方法依旧无法被使用。

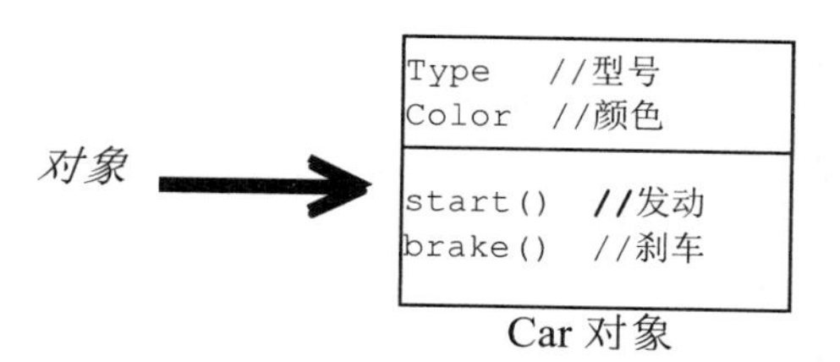

3．连接对象和引用

将新建立的 Car 对象赋值给引用类型变量 car1。这时可以利用引用类型变量 car1 操作对象中的属性和方法了。就像用遥控器控制电视一样。如图 1-9 为对象的创建过程。

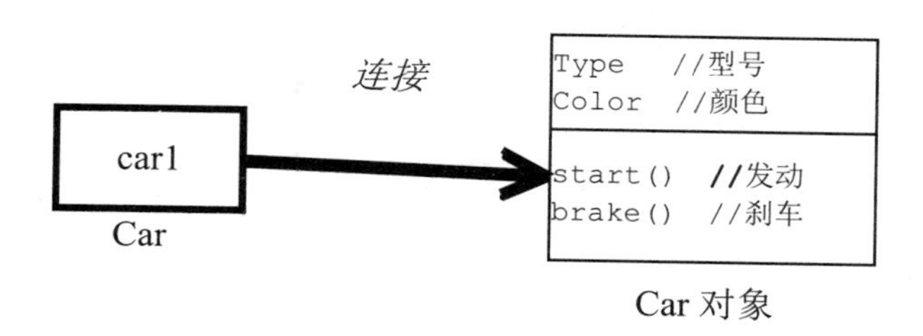

图 1-9　对象的创建过程

在对象为空的情况下，使用对象会出现错误，下面代码中 car1 并没有和一个实例化的对象连接。

```
public class Test {
    public static void main(String[] args) {
            //Car 测试代码
            Car car1 = null;//创建轿车对象
            car1.color="白色";//给 color 属性赋值
            car1.type="轿车";//给 type 属性赋值

            car1.start();//执行汽车发动方法
            car1.brake();//执行踩刹车方法
    }
}
```

运行以后会出现如下错误提示：

```
Problems  @ Javadoc  Declaration  Console
<terminated> Test [Java Application] D:\Program Files\Java\jre8\bin\javaw.exe (2015-12-17 下午3:15:11)
Exception in thread "main" java.lang.NullPointerException
        at chap01.Test.main(Test.java:7)
```

java.lang.NullPointerException 是非常常见的一种错误，叫做空引用(指针)异常。出现这种错误的原因是引用变量 car1 并没有和 Car 对象连接。

任务2 对象的行为

1.2.1 方法的使用

同一个类的每一个对象，都具有相同的属性和方法，只要给予不同属性值，通过方法可以让这些对象呈现出不同的行为。

✧ 需求分析

设计一个类，实现大狗和小狗叫声的不同。

根据需求提取下列特征：

- 狗类的属性：狗的尺寸（size），名字（name）。
- 狗类的方法：吼叫（bark）

通过前面所学知识我们可以如下的方法进行编码：

```
public class Dog1 {
      int size;
      String name;

      public void bark(int barks,String name){
          if(size>10&&size<15){
              System.out.println("我的名字是"+name +", 我要小声叫：汪汪!! ");
          }else if(size>=15){
              System.out.println("我的名字是"+name +", 我要大声叫：汪汪汪汪
                                    汪汪汪汪汪汪汪汪!! ");
          }else{
              System.out.println("这不是狗!! ~~");
          }
      }
}
```

测试类代码，如下：

```
public class Test1 {
    public static void main(String[] args) {
        //测试代码
        Dog1 dog1 = new Dog1();
        dog1.name="小黄";
        dog1.size=12;
        Dog1 dog2 = new Dog1();
        dog2.name="大黄";
```

```
            dog2.size=22;
            Dog1 dog3 = new Dog1();
            dog3.name="黄黄";
            dog3.size=2;

            dog1.bark();
            dog2.bark();
            dog3.bark();
        }
    }
```

运行效果如下：

```
Problems @ Javadoc Declaration Console Progress
我的名字是小黄，我要小声叫：汪汪！！
我的名字是大黄，我要大声叫：汪汪汪汪汪汪汪汪汪汪汪汪！！
这不是狗！！~~
```

这时如果需求改变，要求可以控制狗叫几次，我们可能需要为狗类添加吼叫次数的属性，在方法中添加循环。我们有更好的办法实现这个需求。

✧ 知识准备

1. 带参数的方法

通过带参数的方法可以将狗的吼叫次数直接传递到方法中去。

比如让狗叫 3 声，可以写成 dog1.bark（3）；此时程序在编写的时候可以简洁不少。

首先设置好参数列表，如下所示：

```
public void bark(int barks){   ←——  参数列表（形参）
        for(int i=0 ; i<barks ; i++){
            if(size>10&&size<15){
                System.out.println("我的名字是"+name +"，我要小声叫：汪汪!! ");
            }else if(size>=15){
                System.out.println("我的名字是"+name +"，我要大声叫：汪汪汪
                                汪汪汪汪汪汪汪汪汪!! ");
            }else{
                System.out.println("这不是狗!! ~~");
            }
        }
    }
```

参数列表中的变量表示这个方法可以接受的数据类型，也叫形参。

这时我们在调用 bark（）方法的时候就可以写上我们需要的次数了。这个值会传递给参数列表中的形参 barks。

```
Dog d = new Dog();
d.bark(3);  ←  实参
```

在编写方法时，我们只需要将参数列表中的形参 barks 当成一般的变量使用即可，barks 中的值最终会在方法调用的时候传递过来。

传递

```
d.bark( 3 );            void  bark(      int
                   barks ){
                        ......
                        ......
                        }
```

✧ 编码实施

Dog类：

```
public class Dog2 {
    int size;
    String name;

    public void bark(int barks){
        for(int i=0 ; i<barks ; i++){
            if(size>10&&size<15){
                System.out.println("我的名字是"+name +", 我要小声叫：汪汪!! ");
            }else if(size>=15){
                System.out.println("我的名字是"+name +", 我要大声叫：汪汪汪
                                  汪汪汪汪汪汪汪汪汪!! ");
            }else{
                System.out.println("这不是狗!! ~~");
            }
        }
    }
}
```

测试类：

```
public class Test2 {
    public static void main(String[] args) {
        //测试代码
        Dog2 dog1 = new Dog2();
        dog1.name="小黄";
        dog1.size=12;
        Dog2 dog2 = new Dog2();
        dog2.name="大黄";
        dog2.size=22;

        dog1.bark(2);
        dog2.bark(3);
    }
}
```

执行效果：

```
Problems  @ Javadoc  Declaration  Console
<terminated> Test (1) [Java Application] D:\Program Files\
我是一条小狗，我要小声叫：汪汪！！
我是一条小狗，我要小声叫：汪汪！！
哥是大狗，我要大声叫：汪汪汪汪！！
哥是大狗，我要大声叫：汪汪汪汪！！
哥是大狗，我要大声叫：汪汪汪汪！！
```

✧ 调试运行

1．带参方法使用过程中需注意数据类型要匹配。

Dog 类中 bark 方法的形参为 int 类型。

```
void bark(  int  barks  ){
    ……
    ……
}
```

如果在调用时传递的值不是 int 类型，则会出现如下错误：

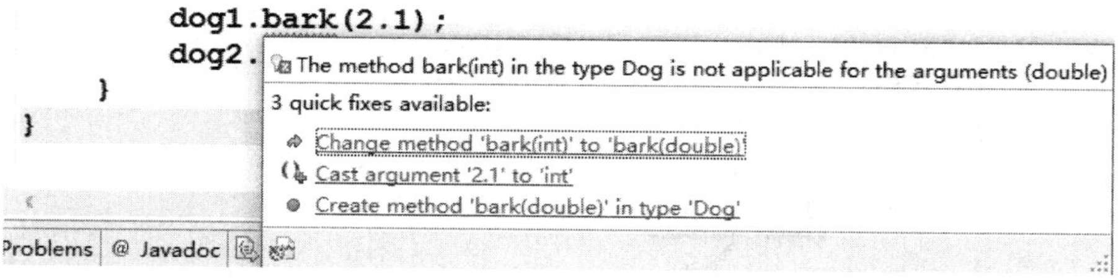

```
dog1.bark(2.1);
dog2.
    }
}
```

The method bark(int) in the type Dog is not applicable for the arguments (double)

3 quick fixes available:

- Change method 'bark(int)' to 'bark(double)'
- Cast argument '2.1' to 'int'
- Create method 'bark(double)' in type 'Dog'

Problems | @ Javadoc

2．带参方法在调用时必须传值。

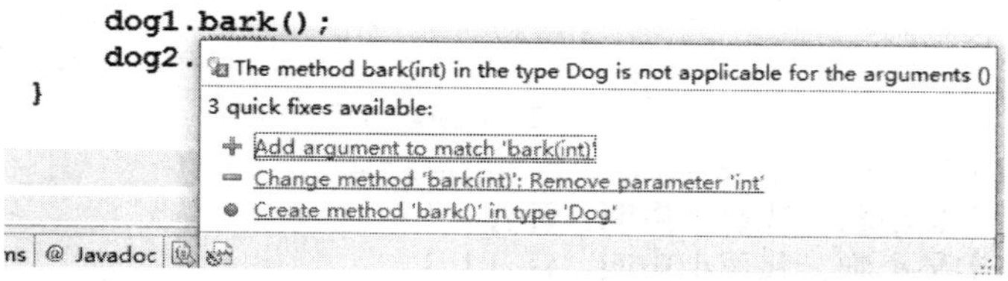

```
dog1.bark();
dog2.
}
```

The method bark(int) in the type Dog is not applicable for the arguments ()

3 quick fixes available:

- Add argument to match 'bark(int)'
- Change method 'bark(int)': Remove parameter 'int'
- Create method 'bark()' in type 'Dog'

ns | @ Javadoc

✧ 维护升级

1．方法的返回值

如果我们还想知道狗具体叫了几次，以便于确定程序是否正常执行。可以通过方法的返回值来实现。需要在 bark（）方法中加入一个计数变量，用于记录狗叫的次数，并将此计数变量作为方法的返回值，返回到调用位置。

Dog 类中的 bark（）方法可以修改成：

```
public  int  bark(int barks){
    int count = 0;
        for(int i=0 ; i<barks ; i++){
        if(size>10&&size<15){
            System.out.println("我是一条小狗，我要小声叫：汪汪!! ");
        }else if(size>=15){
            System.out.println("哥是大狗，我要大声叫：汪汪汪汪!! ");
        }else{
            System.out.println("这是什么狗");
        }
        count++;
    }
    return count;
}
```

方法返回值类型

方法返回值类型

2．定义多个形参

在举例程序中狗缺少名字，我们想给狗起一个名字，并且显示出来。可以重新修改bark（）方法，如下所示：

```
public void bark(int barks  , String name){
for(int i=0 ; i<barks ; i++){
        System.out.println("我的名字是"+name +"，我要小声叫：汪汪!! ");
    }
}
```

多个形参

main方法中的代码：

```
public class Test {
    public static void main(String[] args) {
        //测试代码
        Dog dog1 = new Dog();
        dog1.setSize(15);
        dog1.bark( 3 , "小狗"  );

    }
}
```

多个实参

执行效果：

```
Problems @ Javadoc Declaration Console
<terminated> Test (1) [Java Application] D:\Program File
我的名字是小狗，我要小声叫：汪汪！！
我的名字是小狗，我要小声叫：汪汪！！
我的名字是小狗，我要小声叫：汪汪！！
```

通过这段代码就可以非常方便地传递信息，不过要注意在调用bark（）方法的过程中，实参与形参的数据类型必须对应上，否则程序会出现错误。

1.2.2 构造方法

在前面已经学习过对象的创建过程：

```
Dog dog   =   new  Dog( );
 声明    赋值     创建
```

其中创建过程的Dog( )看起来像调用了一个方法，事实上确实如此。一个类每次创建时都会调用这样的方法，这个方法的名字叫**构造方法**。

✧ 需求分析

构造方法是一个类创建时，必须调用的方法。

```
Dog dog = new Dog( );
```

实际上在创建对象的时候就已经调用了构造方法。

在前面的学习过程中，我们调用方法给属性赋值，比如：

```
Dog dog1 = new Dog();
dog1.setSize(15);
dog1.bark( 3 , "小狗" );
```

但是在前面的程序中并没有定义构造方法，这时编译器会自动创建。当通过 new 关键字创建对象时，构造方法会自动调用，同时也只有在创建对象的时候构造方法才会调用。

通常情况下可以利用构造方法进行对象的初始化操作。有了构造方法上面的例子简化成这样：

```
Dog dog = new Dog(15,"小狗");
dog.bark(3);//吼叫次数
```

执行效果：

```
Problems  @ Javadoc  Declaration  Console
我的名字是大狗，我要大声叫：汪汪汪汪！！
我的名字是大狗，我要大声叫：汪汪汪汪！！
我的名字是大狗，我要大声叫：汪汪汪汪！！
```

可以看出使用构造方法后，我们的赋值操作大大简化，在创建对象的同时可以给属性赋值。

✧ 知识准备

1．构造方法

首先要知道构造方法是创建对象的时候必须执行的。如果我们没写，那么编译器会自动为我们创建，编译器创建出来的构造方法是这样的：

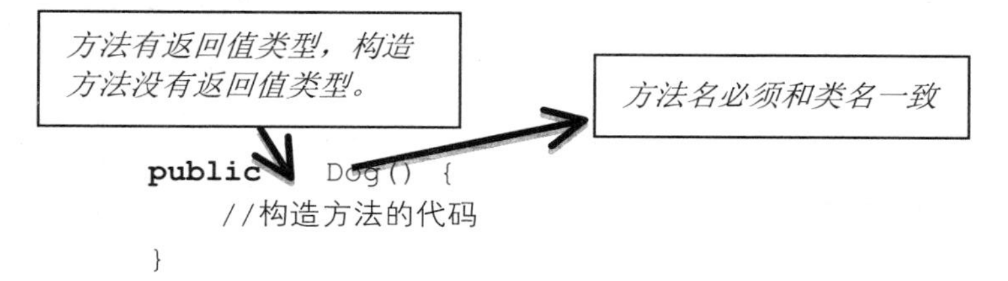

可以看到，这是一个非常特殊的方法，构造方法的两大特点：

（1）构造方法没有返回值类型。

（2）构造方法名必须和类名一样。

2．方法重载

有时方法的功能是相同的（名字可能近似），但是传递的参数是不同的，这些功能相同的方法在使用过程中会遇到很多麻烦；这时我们可以通过方法重载来避免这些麻烦。

实现方法重载的要求：

- 方法名相同；
- 参数列表不同。

✧ 编码实施

1．方法重载

现在编写程序实现两个数字求和，这时参与求和的数据类型是不确定的。现在用一起的思路解决这个问题：

示例代码：

```
public class AddTest {
    public int add1(int a,int b){
        return a+b;
```

```
    }
    public double add2(double a,double b){
        return a+b;
    }

    public static void main(String[] args) {
        AddTest addTest = new AddTest();
        int sum1 = addTest.add1(15, 20);
        double sum2 = addTest.add2(1.5, 1.6);
        System.out.println("整数相加结果: " + sum1);
        System.out.println("小数相加结果: " + sum2);
    }
}
```

执行结果:

Problems @ Javadoc Declaration Console

整数相加结果：35
小数相加结果：3.1

从上例中我们可以看到两个方法都是相加功能，但是计算的数据类型不同，这时得写出两个完全不同的方法，而当我们调用时，就必须记住这两个方法的名字，以防调用错误，否则会给调用带来麻烦。如果使用方法重载的话：

示例代码：

```
public class AddTest {
    public int add(int a,int b){
        return a+b;
    }
    public double add(double a,double b){
        return a+b;
    }

    public static void main(String[] args) {
        AddTest addTest = new AddTest();

        int sum1 = addTest.add(15, 20);
        double sum2 = addTest.add(1.5, 1.6);

        System.out.println("整数相加结果: " + sum1);
        System.out.println("小数相加结果: " + sum2);

    }
}
```

执行结果:

Problems @ Javadoc Declaration Console

整数相加结果：35
小数相加结果：3.1

此时在调用时，仅需要记住一个方法名，通过传递不同的参数 Java 会自动调用不同的方法。

2．构造方法的基本使用

示例代码：

```
public class Dog {
    private int size;
    private String name;                    可以通过构造方法来赋值

    public Dog(int size, String name) {
        if(size>0){
            this.size = size;
        }else{
            System.out.println("对不起你输入的数据不合法");
        }
        this.name = name;
        return size;
    }

    public void setSize(int size) {
        if(size>0){
            this.size = size;
        }else{
            System.out.println("对不起你输入的数据不合法");
        }
    }
    public String getName() {
        return name;
    }
    public void setName(String name) {
        this.name = name;
    }

    public void bark(int barks){
        for(int i=0 ; i<barks ; i++){
            if(size>=15){
              System.out.println("我的名字是"+name +"，我要大声叫：汪汪汪汪汪!! ");
            }else{
              System.out.println("我的名字是"+name +"，我要小声叫：汪汪!! ");
            }
        }
    }
}
```

测试类：

```
public static void main(String[] args) {
        Dog dog = new Dog(13,"大黄");
        dog.bark(3);
}
```

执行结果：

```
Problems  @ Javadoc  Declaration  Console
我的名字是大黄，我要小声叫：汪汪！！
我的名字是大黄，我要小声叫：汪汪！！
我的名字是大黄，我要小声叫：汪汪！！
```

可以看到在测试类中，直接通过构造方法就可以很方便地给 Dog 的属性赋值。

✧ 调试运行

如果类中存在有参构造方法，编译器就不会再添加默认的无参构造方法，从而会导致如图 1-10 的错误：

```
public class Test {
    public static void main(String[] args) {
        Dog dog1 = new Dog(16,"大黄");
        Dog dog2 = new Dog();
                    The constructor Dog() is undefined
                    5 quick fixes available:
                    Add argument to match 'Dog(int)'
                    Add arguments to match 'Dog(int, String)'
                    Change constructor 'Dog(int)': Remove parameter 'int'
                    Change constructor 'Dog(int, String)': Remove parameters 'int, String'
                    Create constructor 'Dog()'
    }
}
```

图 1-10 没有无参构造方法错误

想要避免这个错误，可以在 dog 类中加入一个无参构造方法：

```
public class Dog {
    public Dog(int size, String name) {

        if(size>0){
            this.size = size;
        }else{
            System.out.println("对不起你输入的数据不合法");
        }
        this.name = name;
    }

    public Dog(){            可以加入无参构造方法

    }
    ......
    其他代码略
    ......
}
```

✧ 维护升级

利用构造方法可以更加便捷地为类的属性赋值，而构造方法重载可以提高属性赋值的适应性。如下代码为构造方法重载的示例：

```
public class Dog {
    private int size;
    private String name;

    public Dog(int size, String name) {

        if(size>0){
            this.size = size;
        }else{
            System.out.println("对不起你输入的数据不合法");
        }
        this.name = name;
    }
    public Dog(int size) {

        if(size>0){
            this.size = size;
        }else{
            System.out.println("对不起你输入的数据不合法");
        }
    }

    public int getSize() {
        return size;
    }

    public void setSize(int size) {
        if(size>0){
            this.size = size;
        }else{
            System.out.println("对不起你输入的数据不合法");
        }
    }
    public String getName() {
        return name;
    }
    public void setName(String name) {
        this.name = name;
    }

    public void bark(int barks){
        for(int i=0 ; i<barks ; i++){
            System.out.println("我的名字是"+name +"，汪汪汪汪汪!! ");
```

两个构造方法形成重载

```
        }
    }
}
```

测试类：

```
public static void main(String[] args) {
        Dog dog1 = new Dog(16,"大黄");
        Dog dog2 = new Dog(16);
        dog1.bark(3);
        dog2.bark(3);
}
```

运行效果：

```
Problems @ Javadoc Declaration Console
我的名字是大黄，汪汪汪汪汪！！
我的名字是大黄，汪汪汪汪汪！！
我的名字是大黄，汪汪汪汪汪！！
我的名字是null，汪汪汪汪汪！！
我的名字是null，汪汪汪汪汪！！
我的名字是null，汪汪汪汪汪！！
```

从运行效果中可以看到，dog1 和 dog2 调用了不同的构造方法，所以 dog2 的 name 属性并没有被赋值。

任务 3　高手需要知道的封装：类的封装

前面学习到的代码编写方式其实一直在犯一个低级错误，那就是对于数据的保护，我们之前编写的类的属性是可以直接被访问和修改的：

```
dog1.size = 15;
```

上面是一个正常的数据，看起来使用很方便。不过也可以这样写：

```
dog1.size = -15;
```

这时可能你家的狗就到地平线以下去生活了，很明显这是不符合生活常识的。可悲的是你的程序在实际使用过程中根本没有办法防止这样不合法的数据出现。

✧ 需求分析

针对狗的身高可以为负数的问题，对 Dog 进行重新设计，只接受符合要求的数据。

实现思路：

■ 关闭外界访问和修改属性的途径。

■ 提供一个方法，符合要求的数据才能赋值给属性。

根据以上的思路修改 Dog 之后，就只有特定的方法才能访问和修改属性。

✧ 知识准备

实现封装的基本原则：将属性标记为私有（private），并提供公有（public）的 getter 和

setter 方法。

1．getter 和 setter 方法

通常用于访问和修改属性的方法被称为 getter 和 setter 方法，setter 方法用于设置或修改属性的值，getter 方法用于获取属性的值。

具体写法如下：

```
public int getSize() {
    return size;
}

public void setSize(int size) {
    if(size>0){
        this.size = size;
    }else{
        System.out.println("对不起你输入的数据不合法");
    }
}
```

■ getter 和 setter 方法的命名方式为：set+属性名，get+属性名；并且属性名的首字母大写，比如：setSize ( )，getSize ( )。

虽然不按照这种方法命名，程序也可以正常执行，但是为了提高程序的可维护性和可读性，最好还是按照通用的规矩来。

2．修饰限定符

■ private—私有权限

通过 private 修饰的属性和方法是私有的，不能在类的外部访问，也不能被继承（关于继承会在后续章节提到）。也就是说 private 修饰的属性和方法在该类外部是隐藏的，是看不见的。

■ public—公有权限

与 private 相对的，所有被 public 修饰的属性和方法可以在任何其他类内被访问，不论该类是不是在同一个包。

■ this 关键字

通常 this 关键字表示对象本身。

Dog 类中有如下代码：

```
public void setName(String name) {
        this.name = name;
}
```

表示将接收到的参数赋值给属性。

main 方法中有如下代码：

```
public static void main(String[] args) {
        //测试代码
        Dog dog1 = new Dog();
        dog1.setName("小狗");    // 执行效果等同于 dog1.name=小狗;
        Dog dog2 = new Dog();
        dog2.setName("大狗");   //执行效果等同于 dog2.name=大狗;
}
```

此时可以利用 setName( )方法为 name 属性赋值。当执行 dog1.setName("小狗")时，setName ( ) 中的 this 表示 dog1 的 name 属性；执行 dog2.setName("大狗")时，this 表示 dog2 的 name 属性。

也就是说哪个对象调用的方法，this 就表示那个对象。

✧ 编码实施

下面是一个完整实现封装的例子：

```
public class Dog {
    private int size;
    private String name;

    public int getSize() {
        return size;
    }

    public void setSize(int size) {
        if(size>0){
            this.size = size;
        }else{
            System.out.println("对不起你输入的数据不合法");
        }
    }
    public String getName() {
        return name;
    }
    public void setName(String name) {
        this.name = name;
    }
    void bark(int barks){
        for(int i=0 ; i<barks ; i++){
            if(size>10&&size<15){
                System.out.println("我是一条小狗，我要小声叫：汪汪!! ");
            }else if(size>=15){
                System.out.println("哥是大狗，我要大声叫：汪汪汪汪!! ");
            }else{
                System.out.println("这是什么狗");
            }
        }
    }

}
```

测试类：

```
public class Test {
    public static void main(String[] args) {
        //测试代码
        Dog dog = new Dog();
        dog.setName("大黄");
        dog.setSize(10);
        dog.bark(3);//吼叫次数
    }
}
```

执行效果：

```
Problems @ Javadoc Declaration Console
我的名字是大黄，我要小声叫：汪汪！！
我的名字是大黄，我要小声叫：汪汪！！
我的名字是大黄，我要小声叫：汪汪！！
```

当 Dog 类中的两个属性 size 和 name 的修饰符为 private 的时候，size 和 name 只能在 Dog 类中被访问，此时就可以通过 setSize( )方法过滤属性的值是否合法。而且外界想要给 size 属性赋值，只能通过 setSize( )方法。

在 setter 方法中有这样的语句：this.name = name、this.size = size。等号前后的名字一样，带 this 表示对象本身的属性，不带 this 表示传入的参数，也就是形参。比如 name 是传入的参数(本例中它的值是“小狗”)，this.name 表示 Dog 类的 name 属性,最终将 name 的数据赋值给 this.name。

✧ 调试运行

1．当 Dog 类中的属性设为 private 之后，就无法通过 dog.size 的方式赋值了。此时数据访问更加安全

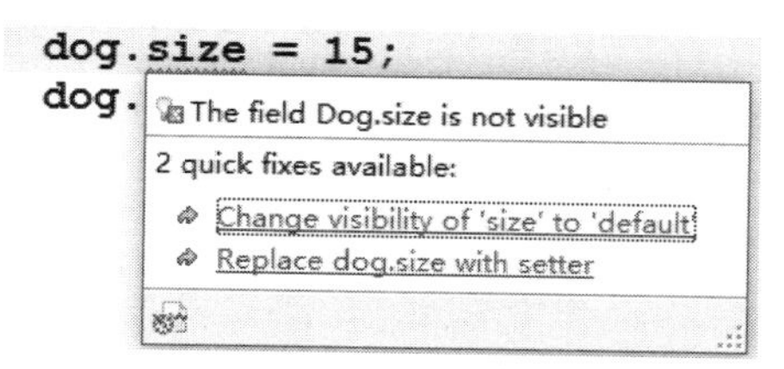
dog.size = 15;
dog.
The field Dog.size is not visible
2 quick fixes available:
Change visibility of 'size' to 'default'
Replace dog.size with setter

2．如果我们给 size 赋值不合法时，程序可以给出错误提示。

测试类中加入如下代码：

```
public class Test {
    public static void main(String[] args) {
        //测试代码
        Dog dog = new Dog();
        dog.setName("大黄");
        dog.setSize(-10);        ← 数据为负数
        dog.bark(3);//吼叫次数
    }
}
```

执行效果：

```
Problems @ Javadoc Declaration Console
对不起你输入的数据不合法
这是什么狗
这是什么狗
这是什么狗
```

✧ 维护升级

在上例中我们还遗留一个问题，现在分析一下setSize()方法：

```
public void setSize(int size) {
        if(size>0){
            this.size = size;
        }else{
            System.out.println("对不起你输入的数据不合法");
        }
}
```

可以看出当dog的size属性赋值为负数时，size属性是不会被赋值的，那么在判断size的时候会不会出现问题？

如果普通变量在没有赋值之前对它判断的话会出现变量未初始化的错误。

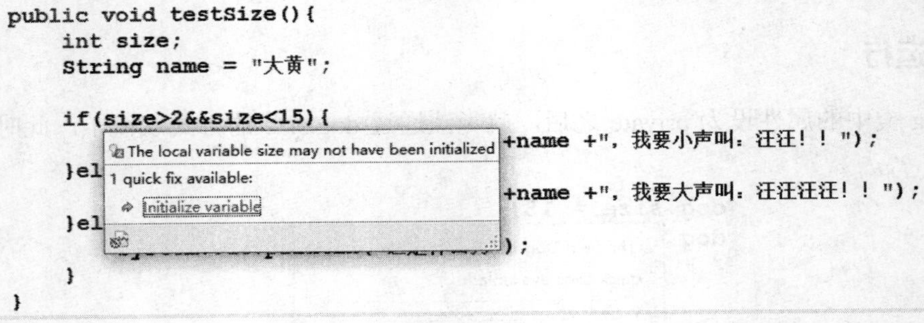

```
public void testSize(){
    int size;
    String name = "大黄";

    if(size>2&&size<15){
        +name +"，我要小声叫：汪汪！！");
    }el
        +name +"，我要大声叫：汪汪汪汪！！");
    }el
        );
    }
}
```

The local variable size may not have been initialized
1 quick fix available:
Initialize variable

在Dog类中，同样也出现这种情况：

看下面这几段代码：

测试类：

```
Dog dog = new Dog();
dog.setSize(-10);
```

传递

Dog类：

```
public void setSize(int size) {
        if(size>0){
            this.size = size;
        }else{
            System.out.println("对不起你输入的数据不合法");
```

执行这里，不会为size赋值

```
        }
}
public void bark(int barks){
        for(int i=0 ; i<barks ; i++){
            if(size>2&&size<15){
            System.out.println("我的名字是"+name +"，我要小声叫：汪汪!! ");
            }else if(size>=15){
                System.out.println("我的名字是"+name +"，我要大声叫：汪汪汪汪!! ");
            }else{
                System.out.println("这是什么狗");
            }
```

执行判断时，size并没有被赋值

```
        }
}
```

当为setSize()方法传递-10时，是没有办法将值赋值给属性size的。此时在bark（）方法中判断size值时，编译器并没有报错。现在我们修改bark()方法输出size的值看看是什么。

修改后的bark()方法：

```
public void bark(){
        System.out.println("size到底是多少？");
        System.out.println("我是：" + size);
    }
```

执行结果：

```
Problems  @ Javadoc  Declaration  Console
对不起你输入的数据不合法
size到底是多少？
我是：0
```

可以看到并没有给size赋值，但是size实际输出的值是0。这是因为属性size是**成员变量**。

成员变量与局部变量的差别：

① 成员变量就是前面我们用到的属性，它声明在类内而不是方法中。

```
public class Dog {
    private int size;
    private String name;
    ......
}
```

② 局部变量是声明在方法中。

```
public class Calculate {
    int a;
    public int add( ){
        int b = 12;
        int sum = a + b;
        return sum;
    }
}
```

③ 成员变量有默认值；而局部变量没有默认值，所以在使用前必须初始化。

```
public class Calculate {
    int a;

    public int add( ){
        int b;
        int sum = a + b;
        return sum;
    }
}
```

局部变量没有初始化，无法编译。

```
The local variable b may not have been initialized
1 quick fix available:
Initialize variable
```

```
public class Calculate {
    int a;
    public int add( ){
        int b = 12;
        int sum = a + b;
```

成员变量没有初始化，但是编译可以通过。

```
        return sum;
    }
}
```

执行效果：

```
Problems @ Javadoc
结果：12
```

可以看出如果成员变量没有赋值的话，它的默认值为0；同时不同数据类型的成员变量，它们的默认值不同。不同数据类型成员变量的默认值如下：

数据类型	默认值
int	0
double	0.0
String	null

项目实训与练习

一、操作题

1．设计一个圆类，属性：半径，方法：计算圆的面积和周长。通过测试类，给定一个半径计算出该圆的面积和周长。

2．创建一个table类，有重量、宽度、长度和高度，计算桌子的面积，显示桌子的数据。

3．看程序写结果：

```
/*动物类*/
class Animal{
    public void eat(){
        System.out.println("吃");
    }
    public void sleep(){
        System.out.println("睡觉");
    }
}
/*兔子类*/
class Rabbit extends Animal{
    public void eat(){
        System.out.println("兔子吃草");
    }
}
/*老虎类*/
class Tigger extends Animal{
    public void eat(){
        System.out.println("老虎吃肉");
    }
```

```
    }
    /*测试类*/
    public class test {
        public void testAnimal(  Animal a  ){
            a.eat();
            a.sleep();
        }
        public static void main(String []agrs){
            test t = new test();
            t.testAnimal(new Rabbit());

        }
    }
```

此程序的运行结果：

二、选择题

1．Java 语言中，方法重载要求（　　）。

A．采用不同的参数列表

B．采用不同的返回值类型

C．调用时用类名或对象名做前缀

D．在参数列表中使用的参数名不同

2．有关 Java 中的类和对象，以下说法错误的是（　　）。

A．同一个类的所有对象都拥有相同的特征和行为

B．类和对象一样，只是说法不同

C．对象是具有属性和行为的实体

D．类规定了对象拥有的特征和行为

3．方法内定义的变量（　　）。

A．一定在方法内所有位置可见

B．可能在方法的局部位置可见

C．在方法外可以使用

D．在方法外可见

4．方法的形参（　　）。

A．可以没有

B．至少有一个

C．必须定义多个形参

D．只能是简单变量

5．return 语句（　　）。

A．不能用来返回对象

B．只可以返回数值

C．方法都必须含有

D．一个方法中可以有多个

三、填空题

1．如果一个方法不返回任何值，则该方法的返回值类型为________。

2．如果子类中的某个方法名、返回值类型和________与父类中的某个方法完全一致，则称子类中的这个方法覆盖了父类的同名方法。

3．一般 Java 程序的类主体由两部分组成：一部分是________，另一部分是________。

4．分别用________关键字来定义类，用________关键字来分配实例存储空间。

5．当一个类的修饰符为________时，说明该类不能被继承，即不能有子类。

项目 2

面向对象的威力

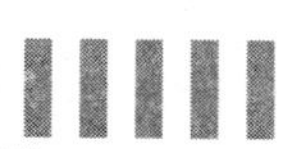

项目目标

通过本项目可以让学生掌握提取父类的方法，学会使用继承和多态；掌握 this、super 和 final 关键字的使用场景和作用。可以利用接口做更深层次的抽象。

项目内容

1. 编写图形变化程序，能够实现不同图形的旋转和放大功能。
2. 为教务处制作一个打印课程表的程序。可以显示班级的名称、学生人数、课程表三项信息。

任务 1　简易多种图形的变化程序：继承的使用

代码重用在编写大型程序时，能够极大地节约编码时间，提升效率。同时绝大多数程序高手都秉承着相同代码致谢一次的信念。无论是代码重用，还是程序高手的信念，继承都是必不可少的。

✧ 需求分析

还记得我们在项目一中图形旋转和放大的例子吗？小明利用面向对象思想以微弱优势战胜了小强。小明设计的程序如图 2-1 所示。

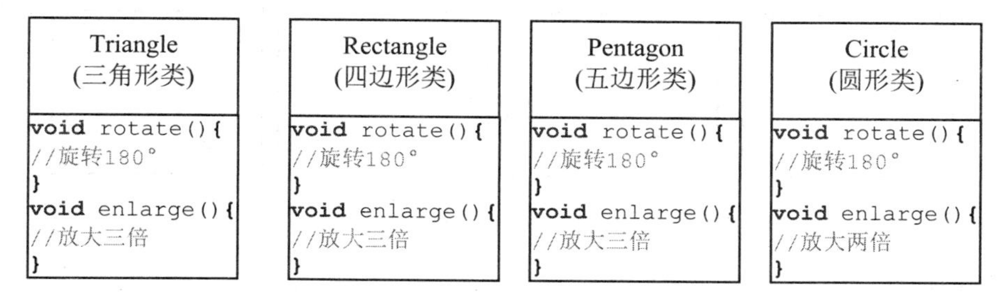

图 2-1　小明设计的程序

在上面的例子中我们可以看到大量的重复代码，我们能不能把相同的部分提取出来呢？然后通过继承让其他的类直接使用呢？这样就能避免重复代码了。

✧ 知识准备

1．技能解析

继承的使用大致可以分为三个步骤：

（1）找出 4 个类中共同的部分。

（2）4 个类都属于形状，并且都有 rotate()和 enlarge()方法，这样我们可以抽取一个新的类 Shape 类。如图 2-2 所示。

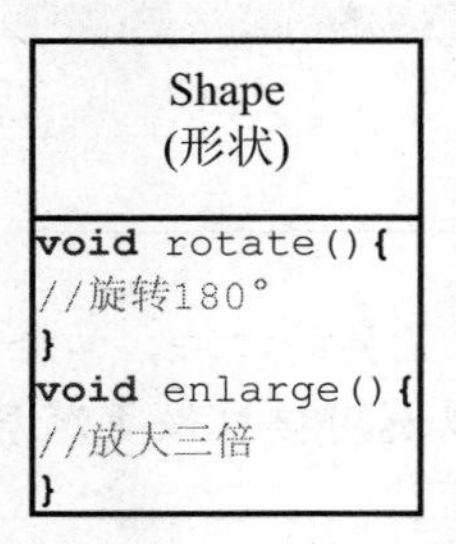

图 2-2 Shape 类

（3）确立 4 个形状和 Shape 类之间的继承关系。如图 2-3 所示。

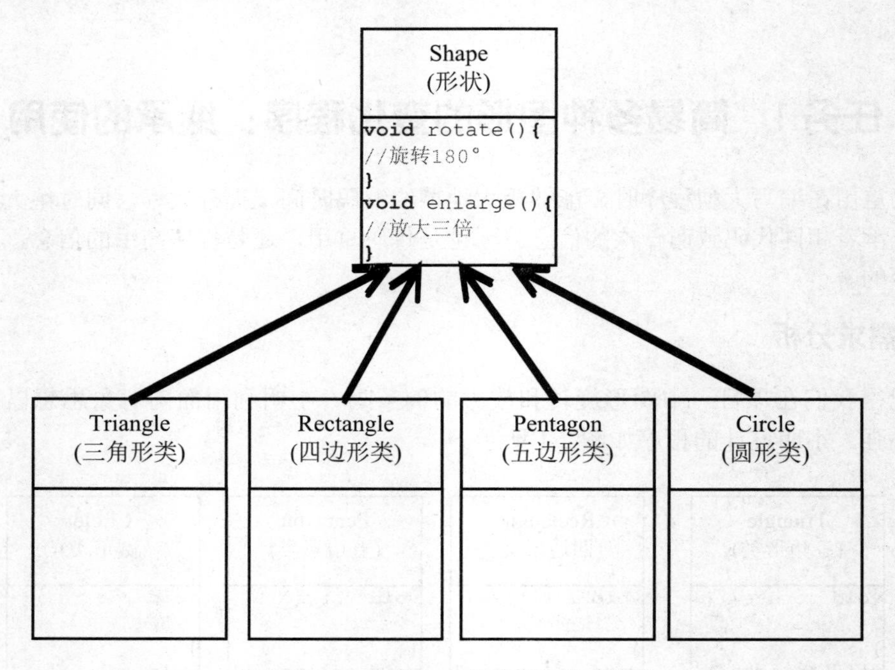

图 2-3 Shape 类与 4 个图形的继承关系

2．知识解析

（1）继承的概念　由一个已有类定义一个新类，称为新类继承了已有类。已有类称为父类，新类称为子类。比如上例中 Shape 类为父类，其余四个图形为子类；可以称为“Triangle 继承 Shape”、“Rectangle 继承 Shape”等。继承以后子类会自动获得父类的功能。但是 private 修饰的属性和方法是不能在子类中直接访问的。子类还可以增加自身特性，定义新的属性和行为，甚至可以重新定义父类中的属性和方法，扩展类的功能。

（2）继承的实现　继承的实现分两个步骤：

■　定义父类

在父类中只定义一些通用的属性与方法，例如：Shape类中的rotate()和enlarge()方法。此步骤跟之前学习过的类的定义没有本质区别。

■　定义子类

子类定义格式：

```
[类修饰符] class 子类名 extends 父类名{
    //代码
}
```

注意：一个子类只能继承一个父类，即**单继承**。但一个父类可以有多个子类。

✧ 编码实施

（1）没有特殊需求的 **Triangle** 类和 **Rectangle** 类的继承写法。

Shape 类：

```
public class Shape {
    //旋转
    public void rotate(){
        System.out.println("旋转180度");
    }
    //放大
    public void enlarge(){
        System.out.println("放大三倍");
    }
}
```

Triangle 类：

```
public class Triangle extends Shape {

}
```

Rectangle 类：

```
public class Rectangle extends Shape {

}
```

测试类：

```
public static void main(String[] args) {
        Triangle triangle = new Triangle();
        triangle.rotate();

        Rectangle rectangle = new Rectangle();
        rectangle.rotate();

}
```

执行效果：

```
Problems @ Javadoc Declaration Console
旋转180度
旋转180度
```

可以看出在 **Triangle** 类和 **Rectangle** 类中并没有写任何内容，但是我们一样可以执行 rotate()方法，并且能够得到正确的执行结果。这就是继承的妙用，它大大提高了代码重用的效率。

➢ 练一练：

仿照上面的过程，创建其他两个子类对象，并测试运行。

不过其中有一个 Circle 类（圆类）比较特殊，它要求放大两倍，并且要围绕正文旋转。那么 Circle 类继承 Shape 类后是无法完成需求的，此时可以利用继承的另一个特性——**重写**。所谓方法重写就是子类定义的方法和父类的方法具有**相同的名称、参数列表、返回类型和访问修饰符**。

（2）需要特殊处理的 **Circle** 类的继承写法。

Circle 类：

```
public class Circle extends Shape {
    public void rotate() {   //方法重写
        System.out.println("绕轴旋转");
    }
    public void enlarge() {  //方法重写
        System.out.println("放大两倍");
    }
}
```

测试类：

```
public static void main(String[] args) {
        Circle circle = new Circle();
        circle.rotate();
}
```

（3）**执行结果：**

```
Problems @ Javadoc Declaration Console
绕轴旋转
```

继承是非常灵活的，从父类集成过来的方法满足需求就可以在子类中直接使用；而不满足需求的话，在子类中可以通过重写，对继承过来的方法进行修改，这种修改对父类和其他子类是没有任何影响的。

另外方法重写会覆盖父类的同名方法。也就是说在 Circle 类中调用 rotate()方法，将会是子类中重写的新方法，而不是父类中的 rotate()方法。子类的属性与父类的属性相同时，也会出现覆盖的现象。

✧ 调试运行

在父类中并不是所有的属性和方法都可以被子类继承，在我们设计父类的过程中可以通过 private 和 public 来区分哪些可以继承，那些不可以继承。

（1）**Shape 类：**

```
public class Shape {

    private String name;

    private void setName(String name){
        this.name = name;
    }
    //旋转
    public void rotate(){
        System.out.println("旋转 180 度");
    }
    //放大
    public void enlarge(){
        System.out.println("放大三倍");
    }
}
```

name 属性和 setName()方法均被 private 修饰。

（2）**Circle 类**正常的继承 Shape 类：

```
public class Circle extends Shape {
    public void rotate() {
        System.out.println("绕轴旋转");
    }
    public void enlarge() {
        System.out.println("放大两倍");
    }
}
```

（3）测试类出现了两个错误：

```
 5    public static void main(String[] args) {
 6
 7        Circle circle = new Circle();
 8        circle.rotate();
 9        circle.name = "圆形";
10        circle.setName("圆形");
11    }
```

第 9 行错误：Shape.name 是不可见的（证明父类 Shape 中的 name 是无法使用的）。

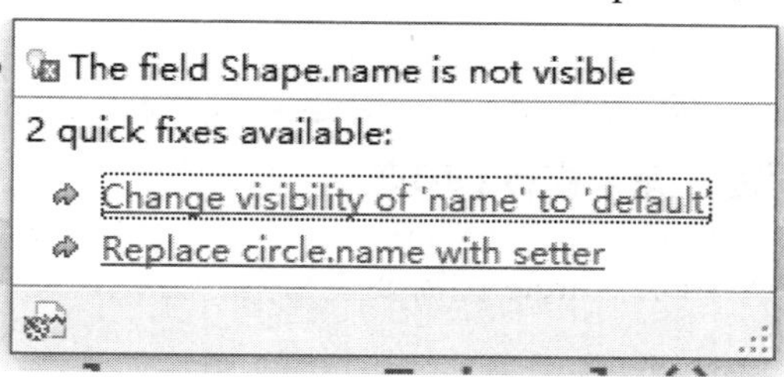

The field Shape.name is not visible
2 quick fixes available:
Change visibility of 'name' to 'default'
Replace circle.name with setter

第10行错误:Shape中的setName(String)方法是不可见的。

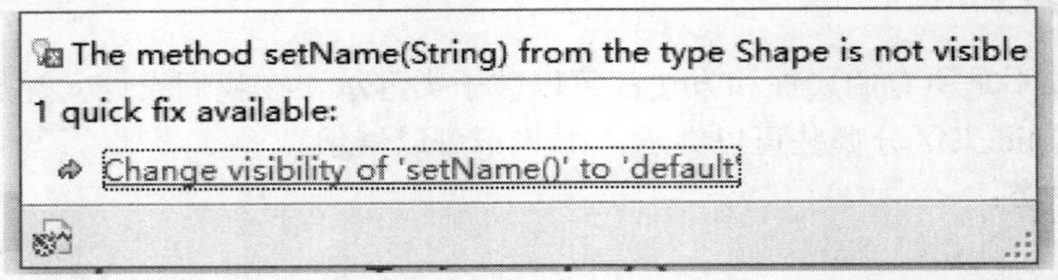

```
The method setName(String) from the type Shape is not visible
1 quick fix available:
Change visibility of 'setName()' to 'default'
```

从上面的试验中我们可以看到，private 类型的数据是无法被子类继承的。我们可以把Shape类中setName(String)方法的修饰符改成public。

修改后的Shape类：

```
public class Shape {
    private String name;
    public void setName(String name){
        this.name = name;
    }
    public String getName() {
        return name;
    }
    //旋转
    public void rotate(){
        System.out.println("我是"+name+"旋转180度");
    }
    //放大
    public void enlarge(){
        System.out.println("我是"+name+"放大三倍");
    }
}
```

Triangle类:

```
public class Triangle extends Shape {

}
```

测试类：

```
5     public static void main(String[] args) {
6
7         Triangle triangle = new Triangle();
8         triangle.setName("三角形");
9         triangle.rotate();
10    }
```

执行结果：

```
Problems  @ Javadoc  Declaration  Console
我是三角形旋转180度
```

而Circle类的情况有所不同，Circle类需要重写两个方法，此时它是不能直接使用name属性的。

Circle 类:

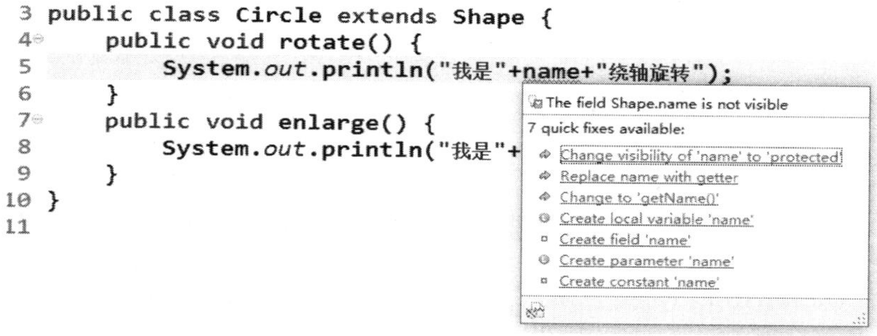

```
3 public class Circle extends Shape {
4⊖     public void rotate() {
5         System.out.println("我是"+name+"绕轴旋转");
6     }
7⊖     public void enlarge() {
8         System.out.println("我是"+
9     }
10 }
11
```

The field Shape.name is not visible
7 quick fixes available:
Change visibility of 'name' to 'protected'
Replace name with getter
Change to 'getName()'
Create local variable 'name'
Create field 'name'
Create parameter 'name'
Create constant 'name'

此时需要使用 Shape 类中提供的 getter 方法才能访问 name 属性。

修改后的 Circle 类:

```
public class Circle extends Shape {
    public void rotate() {
        System.out.println("我是"+getName()+"绕轴旋转");
    }
    public void enlarge() {
        System.out.println("我是"+getName()+"放大两倍");
    }
}
```

测试类:

```
    public static void main(String[] args) {
        Circle circle = new Circle();
        circle.setName("圆形");
        circle.rotate();
    }
```

执行结果:

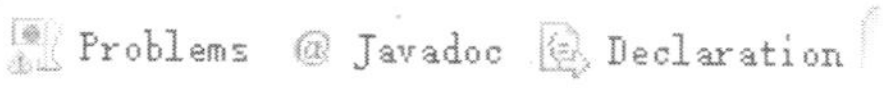

Problems @ Javadoc Declaration

我是圆形绕轴旋转

✧ 维护升级

利用构造方法可以更快捷地对类进行初始化，但是子类是不能继承父类构造方法的，因为父类的构造方法用来创建父类对象，子类需定义自己的构造方法，创建子类对象。子类的构造方法中，通过 super 关键字调用父类的构造方法。

现在将上例修改，利用构造方法进行属性的赋值：

```
public class Shape {
    private String  name;
    public Shape(String name) {
        this.name = name;    //this 表示当前对象
        System.out.println("有参构造方法被调用");
    }
    public Shape(){
        System.out.println("无参构造方法被调用");
    }
```

```
    public void rotate(){
        System.out.println("我是"+name+"旋转180度");
    }
    public void enlarge(){
        System.out.println("我是"+name+"放大三倍");
    }
}
```

我们添加了两个构造方法来提升属性初始化的效率。下面是子类和测试类的相关代码：

Triangle 类

```
public class Triangle extends Shape {

}
```

测试类：

```
5    public static void main(String[] args) {
6
7        Triangle triangle = new Triangle();
8        triangle.rotate();
9    }
```

不传递参数时没有问题

如果想要通过构造方法给name赋值时，会提示“构造方法没有被定义”的错误。

```
5    public static void main(String[] args) {
6
7        Triangle triangle = new Triangle("三角形");
8        triangle.rotate();
9    }
10
11 }
12
```

The constructor Triangle(String) is undefined
2 quick fixes available:
Remove argument to match 'Triangle()'
Create constructor 'Triangle(String)'

现在我们修改**Triangle**类，为**Triangle**类加入构造方法，以便为name属性赋值。如果像下图一样直接赋值是不行的，因为name属性是父类中的私有成员变量，不能被子类直接访问。

```
5    public Triangle(String name){
6        this.name = name;
7    }
8 }
9
```

The field Shape.name is not visible
3 quick fixes available:
Change visibility of 'name' to 'protected'
Create getter and setter for 'name'...
Create field 'name' in type 'Triangle'

如果要对name属性赋值，只能通过父类的构造方法才行，所以**Triangle**类正确的写法应该是在自己的构造方法中，利用super关键字调用父类构造方法。下面是示例代码：

Triangle 类：

```
public class Triangle extends Shape {
    public Triangle(String name) {
        super(name);    //调用父类的有参构造方法
    }
```

```
    public Triangle() {
            super();          //调用父类的无参构造方法
        }
    }
```

测试类:

```
    public static void main(String[] args) {
            Triangle triangle1 = new Triangle("三角形");
            Triangle triangle2 = new Triangle();
    }
```

执行结果:

Problems @ Javadoc Declaration Console

有参构造方法被调用
无参构造方法被调用

注意：super([参数]);必须是子类构造方法的第一条语句。如果该语句省略，则会自动调用父类无参构造方法。因为子类创建对象时，先创建父类对象，再创建子类对象。

同样当出现覆盖现象时。可以通过 super 关键字调用父类的方法：

比如在 Circle 类中可以利用 super 关键字调用父类的方法：

```
    super. rotate() ;    //super 指父类对象
```

如果用 this 关键字，则表示调用当前对象的 rotate()方法：

```
    this. rotate();      //this 指当前对象，通常情况下是可以省略的
```

通过使用 super 关键字与 this 关键字，可以很显式地区分开调用的是当前对象的成员，还是父对象的成员。

在父类的构造方法中也用到了 this：

```
    public Shape(String name) {
            // this.name 表示当前对象的属性
    this.name = name;
        }
```

在形式参数和属性名相同时，需要通过 this 来区分参数和属性

方法体中定义的变量，或方法的形式参数与对象的成员变量名相同时，也必须用 this 关键字指明当前对象的成员变量。

任务 2 强化图形变化程序：多态的使用

多态是面向对象中非常重要的特性之一，随着学习的深入多态也是无处不在的，通过多态我们可以更加灵活、方便地编写程序。

✧ 需求分析

我们曾经做过的变形程序扩展性还不够强大，如果想要加入新图形，继承 Shape 类以后，还需要在测试类中创建新图形的对象和引用，然后才能执行变形方法。例如：

```
    Oval oval = new Oval();
    oval.rotate();
```

虽然这些图形调用的 rotate()方法是一样的，但是我们不得不利用不同的对象去调用它；

这种情况会导致今后在简化代码、统一处理代码的时候带来很多麻烦。如果用同一个对象名调用同一个方法，在实际执行的时候，却可以根据实际情况得到不同的执行结果。看起来很神奇，不过利用多态可以很轻易地实现。

现在我们想要建立一个形状工厂，用户只要将形状告诉形状工厂，形状工厂就可以将该形状放大和旋转。

测试类代码：

```
public static void main(String[] args) {
        ShapeFactory factory = new ShapeFactory();
        factory.show( new Circle("圆形") );
        factory.show( new Triangle("三角形") );
}
```

执行结果：

Problems @ Javadoc Declaration Console

我是圆形绕轴旋转
我是圆形放大两倍
我是三角形旋转180度
我是三角形放大三倍

✧ 知识准备

1．技能解析

想要了解多态，先从创建对象的过程看起，一个对象如果想要正常使用通常来说要经历三步：

```
Circle circle  =  new  Circle ();
```

①（Circle circle） ③（=） ②（new Circle ()）

（1）声明引用。

（2）创建对象。

（3）连接对象和引用。

在前面学习过程中已经了解到，引用类型与对象类型必须相符，此例中两者皆为Oval。但是在多态下，应用与对象是可以有不同类型的。

```
Shape shape  =  new Circle();
```

Shape和Circle不属于同一类型，引用变量的类型是Shape，对象的类型为Circle。运用多态时，引用类型可以是实际对象类型的父类。换句话说，子类对象可以赋值给父类引用。

示例：

```
public static void main(String[] args) {

        Shape shape = new Circle("圆形");
        shape.rotate();

        shape = new Triangle("三角形");
        shape.rotate();
}
```

执行结果：

可以看到调用方法时使用的都是 shape.rotate()，但是得到了不同的结果。两次调用引用和方法的名字虽然是一样的，但是实际调用的方法却是不同的。

第一次执行 shape.rotate();时：

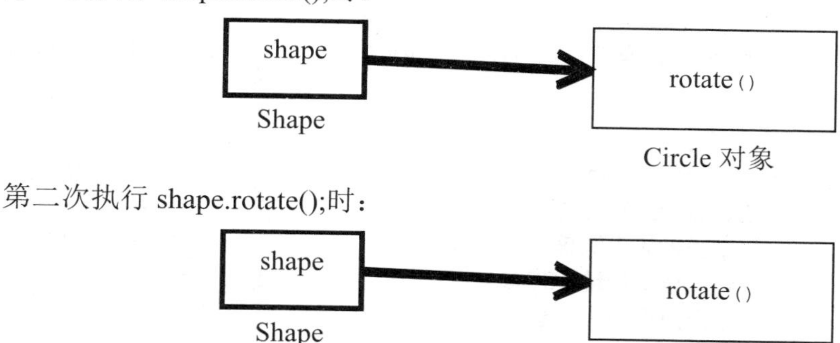

第二次执行 shape.rotate();时：

shape

Shape

rotate ()

Triangle 对象

同样在参数和返回值中也可以实现多态。

示例代码：

```
public class ShapeFactory {

    public void show(Shape shape){
        shape.rotate();
        shape.enlarge();
    }

}
```

shape 参数可以接收任何 Shape 类型的子类对象，并且会根据传入的实际对象，调用对应的方法。

Rectangle　***Oval***　***Circle***　***Triangle***

2．知识解析

多态分为静态多态和动态多态。静态多态指通过同一个类中的方法重载实现的多态，动态多态指通过类与类之间的方法重写实现的多态。

静态多态是编译时多态，指程序会根据参数的不同来调用相应的方法，具体调用哪个被重载的方法，由编译器在编译阶段静态决定。

动态多态是运行时多态，指在运行时根据调用该方法的实例的类型来决定调用哪个重写方法。

静态多态对程序运行更有效率，动态多态更具有灵活性。

注意：如果子类重写了父类的方法，则调用子类的方法；如果子类未重写父类的方法，则调用父类的方法。因此，父类对象可以通过引用子类的实例调用子类的方法。

✧ 编码实施

（1）创建 ShapeFactory 类，此类中包含 show()方法，能够实现变形操作：

```
public class ShapeFactory {
    public void show(Shape shape){
        shape.rotate();
        shape.enlarge();
    }
}
```

（2）利用 ShapeFactory 显示编写结果：

```
public static void main(String[] args) {
        ShapeFactory factory = new ShapeFactory();
        factory.show( new Circle("圆形") );
        factory.show( new Triangle("三角形") );
}
```

（3）执行结果：

```
Problems @ Javadoc Declaration Console
我是圆形绕轴旋转
我是圆形放大两倍
我是三角形旋转180度
我是三角形放大三倍
```

如果需要对这个程序进行扩展，加入新图形，一样非常方便，对程序的主体结构几乎没有影响。比如新加入一个椭圆类（Oval），旋转的需求不变，放大 5 倍，仅仅需要创建一个 Oval 类，根据需求的变化重写父类方法即可：

```
public class Oval extends Shape {
    public Oval(String name) {
        super(name);
    }
    public void enlarge() {
        System.out.println("我是"+getName()+"放大五倍");
    }
}
```

接着修改测试类：

```
public static void main(String[] args) {
        ShapeFactory factory = new ShapeFactory();
        factory.show( new Circle("圆形") );
        factory.show( new Triangle("三角形") );
        factory.show( new Oval("椭圆") );
}
```

执行效果：

```
Problems @ Javadoc Declaration Console
我是圆形绕轴旋转
我是圆形放大两倍
我是三角形旋转180度
我是三角形放大三倍
我是椭圆旋转180度
我是椭圆放大五倍
```

✧ 调试运行

练一练：

下面关于多态定义有哪些错误？

父类 Shape 中有如下方法：

```
public void rotate(){
        System.out.println("我是"+name+"旋转 180 度");
}
```

子类 Cirle 类中有如下方法：

```
public void rotate(String name){
        System.out.println("我是"+name+"旋转 180 度");
}
```

提示：有一个错误。

✧ 维护升级

在前面的学习过程中，子类和父类之间的关系，类似于新版本程序和老版本程序之间的关系。子类就是新版本的父类，子类在某种程度上对父类进行了修改。不过在 java 中有一些类是用来完成某种标准功能的类，例如，系统类 String、Byte 和 Double，或定义已经很完美，不需要生成子类的类，也就是说这样的类是不应该被继承，形成新版本的。此时需要利用 final 关键字，将这样的类定义为最终类。

最终类的声明格式：

```
final  class  类名{
......
}
```

比如 ShapeFactory 类写的很完美了，不希望别人继承、修改，就可以将 ShapeFactory 定义成最终类。

示例代码:

```
public final class ShapeFactory {
    public void show(Shape shape){
        shape.rotate();
        shape.enlarge();
    }
}
```

此时 ShapeFactory 将不能被继承和修改。

注意：final 类不能被继承，没有子类，final 类中的方法默认是 final 的。

同样 final 关键字还可以用在方法和变量的修饰上。

final 用来修饰方法，则表示把方法锁定，以防任何继承类修改它的含义。也就是说 final 方法不能被子类的方法覆盖，但可以被继承。

注意：父类的 private 成员方法是不能被子类方法覆盖的，因此 private 类型的方法默认是 final 类型的，而且 final 不能用于修饰构造方法。

Shape类中的方法前加入final:

```
public final void rotate(){
        System.out.println("我是"+name+"旋转180度");
}
```

则在Circle类中会出现错误:不能重写Shape类的final方法。

```
9⊖    public void rotate() {
10        System.o                              )+"绕轴旋转");
11    }
12⊖    public void
13        System.o                              )+"放大两倍");
14    }
```

```
Cannot override the final method from Shape
1 quick fix available:
Remove 'final' modifier of 'Shape.rotate'(..)
```

final修饰的变量就变成了常量，只能被赋值一次，赋值后值将不再改变。

final修饰的变量:

```
5     private final String name = "形状";
6
7⊖    public Shape(String name) {
8         this.name = name;
9     }
10 //     Syste                                 调用
11
12⊖    public St
```

```
The final field Shape.name cannot be assigned
1 quick fix available:
Remove 'final' modifier of 'name'
```

在构造方法中为name重新赋值，会提示错误。

任务3　课表打印程序：抽象与接口

✧ 需求分析

（1）所有的类都应该实例化吗？

在前面的学习过程中知道**创建对象可以这样写**:

```
Circle circle = new Circle ();
```

也可以这样写:

```
Shape shape = new Circle();
```

可是如果这样写就会觉得很别扭:

```
Shape shape = new Shape();
```

创建了一个形状对象，到底是什么形状，会让我们很困惑，这就意味着有些类并不适合实例化，甚至不能被实例化。利用抽象或接口都能够解决这个问题。

（2）如何更好地设计程序结构，提升团队合作效率。

软件工作室接到了一个新的任务：每年开学每个班级都需要打印课程表，信息工程系希望为教务处制作一个打印课程表的程序。可以显示班级的名称、学生人数、课程表三项信息。现在以信息工程系的班级为实验对象，制作这套打印程序。

注意：在此例中所有班级的课表均不同。

执行效果：

```
Problems  @ Javadoc  Declaration  Console  Progress  LogCat
周一课表：
12节            34节            56节            78节
组装维护         SQLServer       JAVA            体育
---------------------------------------------------
软件班，人数：15
周一课表：
12节            34节            56节            78节
HTML            JAVA            思想品德         体育
---------------------------------------------------
软件专业，张三老师
周一课表：
12节            34节            56节            78节
JSP                             office
```

✧ 知识准备

1．技能解析

经过初步分析所需的类有信息工程系所有班级的类、教务处类。其中班级以“计算机应用”和“软件”为例。教务类可以打印所有班级的课程表，它们之间的关系如图 2-4 所示。

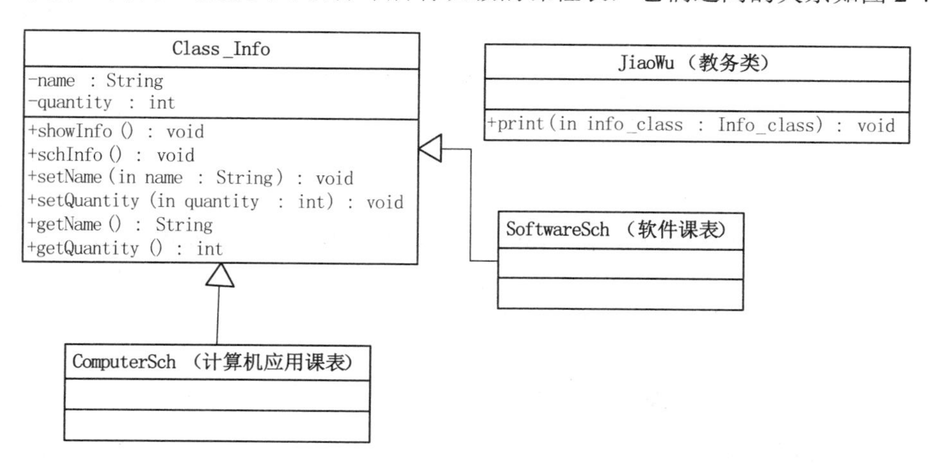

图 2-4　类关系图

Class_Info 类为父类，其中 showInfo()方法用于显示班级信息，schInfo()方法用于显示课表信息，由 SoftwareSch 类和 ComputerSch 类去继承它；最后在 JiaoWu 类中通过 print()方法将课表打印出来。其中我们会遇到一个问题：在 Class_Info 类中 schInfo( )方法用于显示课表信息，该方法要求必须用子类的新版本“覆盖”掉。因为每个班级的课表是不一样的。

2．知识解析

本任务中主要涉及到 abstract 关键字和 interface 关键字，分别是抽象和接口。

（1）**abstract 关键字**可以修饰类和方法，分别表示抽象类、抽象方法。抽象类是可以被子类继承却**不能创建实例**的类。抽象类中可以声明只有方法头没有方法体的**抽象方法**，方法体由子类实现。

抽象类和抽象方法的编写格式：

```
[public] abstract class 类名{
[public] abstract 方法返回值 方法名();
//抽象方法没有方法体
}
```

（2）**interface 关键字**只能修饰类，被 interface 修饰的类称为接口，这是一种特殊的类，只包含常量和抽象方法。一般只表示一种“规范”。例如手机上的 microUsb 接口能够完成充电、数据传输等功能，几乎所有的手机厂商都遵循该规范，这样的好处就是随便找到一个充电器就能用。

接口的编写格式：

```
[public] interface 类名{
//接口中的方法都是抽象方法
}
```

如果某一个类实现了一个接口，则必须实现该接口中的所有方法。

✧ 编码实施

1．修改变形程序防止 Shape 类实例化。只需在 class 前加入 abstract。

```
public abstract class Shape {
    代码略……
}
```

此时如果对 Shape 实例化就会提示错误：

```
7⊖    public static void main(String[] args) {
8         Shape shape = new Shape();
9                       ⊗Cannot instantiate the type Shape
10    }
11
```

2．为不同班级打印课表。

通过上面的分析可知，父类 Class_Info 中有两个关键方法，分别用于显示班级信息和课表信息。其中显示班级信息的方法是可以让子类继承并使用的，但是显示课表信息的方法对于每个班级来说是不同的，需要子类去覆盖。示例代码如下：

Class_Info 类：

抽象类（无法被实例化）

```
public abstract class ClassInfo {
    private String name;     //班级名
    private int quantity;    //人数

    /**
     * 显示班级信息
     */
    public void showInfo(){
        System.out.println(name + "班，人数：" + quantity);
    }
    /**
     * 显示课表信息
     */
    public abstract void schInfo();
```

抽象方法，没有方法体，由子类给出具体实现

```
        public String getName() {
            return name;
        }
        public void setName(String name) {
            this.name = name;
        }
        public int getQuantity() {
            return quantity;
        }
        public void setQuantity(int quantity) {
            this.quantity = quantity;
        }
    }
```

ComputerSch 类：

```
    public class ComputerSch extends ClassInfo{
        public void schInfo() {
            System.out.println("周一课表：");
            System.out.println("12 节\t\t34 节\t\t56 节\t\t78 节");
            System.out.println("组装维护\t\tSQLServer\tJAVA\t\t 体育");
        }
    }
```

SoftwareSch 类：

```
    public class SoftwareSch extends ClassInfo{
        public void schInfo() {
            System.out.println("周一课表：");
            System.out.println("12 节\t\t34 节\t\t56 节\t\t78 节");
            System.out.println("HTML\t\tJAVA\t\t 思想品德\t\t 体育");
        }
    }
```

JiaoWu 类：

```
    public class JiaoWu {
        public void print(ClassInfo classInfo){

    System.out.println("-------------------------------------------------------
");
            classInfo.classInfo();
            classInfo.schInfo();
        }
    }
```

测试类：

```
    public static void main(String[] args) {
            JiaoWu jiaoWu = new JiaoWu();
            ComputerSch computerSch = new ComputerSch();
            computerSch.setName("计算机应用");
            computerSch.setQuantity(20);
            jiaoWu.print(computerSch);
```

```
        SoftwareSch softwareSch = new SoftwareSch();
        softwareSch.setName("软件");
        softwareSch.setQuantity(15);
        jiaoWu.print(softwareSch);
    }
```

执行结果：

```
Problems @ Javadoc Declaration Console Progress LogCat
--------------------------------------------------
计算机应用班，人数：20
周一课表：
12节              34节              56节              78节
组装维护          SQLServer         JAVA              体育
--------------------------------------------------
软件班，人数：15
周一课表：
12节              34节              56节              78节
HTML              JAVA              思想品德          体育
```

✧ 调试运行

1．抽象方法只能在抽象类中出现。如果一个类不是抽象类，却包含了抽象方法，则会在方法和类上出现如下错误：

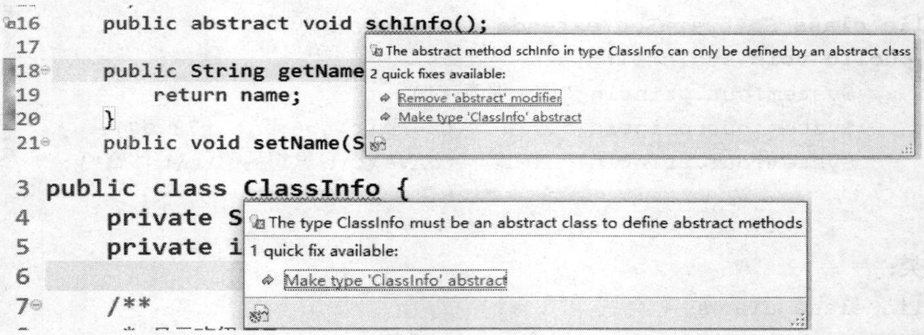

```
16     public abstract void schInfo();
17
18     public String getName
19         return name;
20     }
21     public void setName(S
```

```
The abstract method schInfo in type ClassInfo can only be defined by an abstract class
2 quick fixes available:
  Remove 'abstract' modifier
  Make type 'ClassInfo' abstract
```

```
3 public class ClassInfo {
4     private S
5     private i
6
7     /**
```

```
The type ClassInfo must be an abstract class to define abstract methods
1 quick fix available:
  Make type 'ClassInfo' abstract
```

2．子类必须实现父类的抽象方法。如果继承了一个抽象类，而没有实现它的抽象方法时，会提示如下错误：

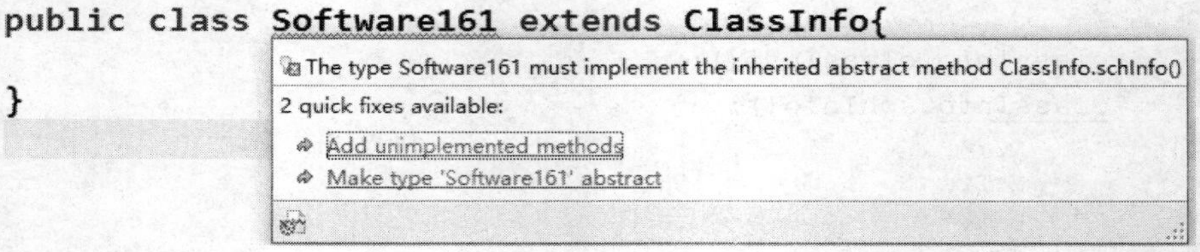

```
public class Software161 extends ClassInfo{

}
```

```
The type Software161 must implement the inherited abstract method ClassInfo.schInfo()
2 quick fixes available:
  Add unimplemented methods
  Make type 'Software161' abstract
```

✧ 维护升级

如果现在需求改变了，教务类不仅要打印班级课表，还需要打印教师课表，这时原来类的结构设计明显不符合需求了，那么现在需要重新设计类的结构。可是现在有教师类和班级信息类，我们该把提取出来的公共部分放到什么类中呢？

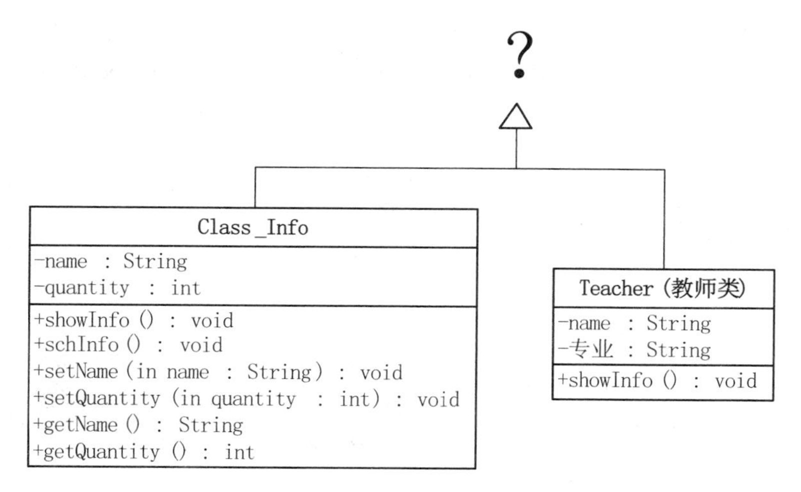

可以感觉到现在代码的扩展性受到了限制，按照原来的思路设计也会感觉到很困惑。这时需要一种方法，既能够避免这种困惑，又可以使教务类继续利用多态的方式显示课表。

真正需要的方法是：

（1）一种确保所有“涉及到课表显示的类”都有相同规范的方法。

（2）一种可以运用到多态的方法。

（3）这种方法可以让“课表显示”功能应用在任意需要该功能的类中（比如：班级类、教师类、实训类等）。

这种方法就是接口，利用接口定义“课表显示”功能的规范，然后所有需要该功能的类只要遵循这个规范，就可以实现多态带来的一系列好处。利用接口重新设计的类关系图如图 2-5 所示。

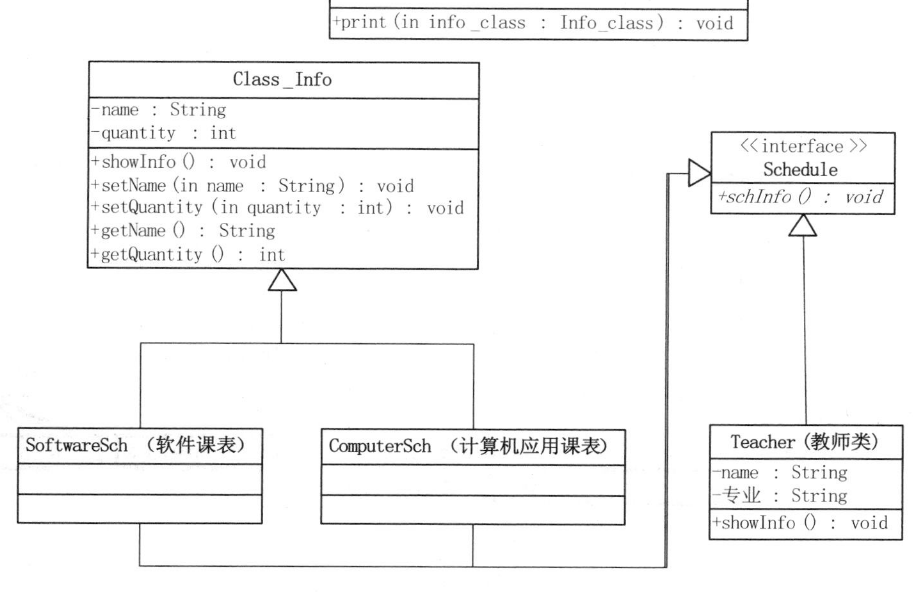

图 2-5　类关系图

将课表显示功能提取到接口中，想要使用课表显示功能，就要实现 Schedule 接口。SoftwareSch 类和 ComputerSch 类不但继承了 Class_Info，还实现了 Schedule 接口，而 Teacher 类也实现了 Schedule 接口。

示例代码：

Schedule 接口：

```
public interface Schedule {
    /**
     * 显示课表信息
     */
    public void schInfo();
}
```

接口的方法，接口中的方法均为抽象方法，无需用 abstract 关键字指定

ClassInfo 类：

```
public class ClassInfo {
    private String name;     //班级名
    private int quantity;    //人数

    /**
     * 显示班级信息
     */
    public void classInfo(){
        System.out.println(name + "班，人数：" + quantity);
    }

    public String getName() {
        return name;
    }
    public void setName(String name) {
        this.name = name;
    }
    public int getQuantity() {
        return quantity;
    }
    public void setQuantity(int quantity) {
        this.quantity = quantity;
    }
}
```

实现接口的关键字，类似于继承的 extends

接口的名字

ComputerSch 类：

```
public class ComputerSch extends ClassInfo implements Schedule {
public void schInfo()
{
        classInfo();     //显示班级信息
        System.out.println("周一课表：");
        System.out.println("12 节\t\t34 节\t\t56 节\t\t78 节");
```

实现接口的方法

```
        System.out.println("组装维护\t\tSQLServer\tJAVA\t\t 体育");
    }
}
```

SoftwareSch 类:

```
public class SoftwareSch extends ClassInfo implements Schedule{
    public void schInfo() {
        classInfo();    //显示班级信息
        System.out.println("周一课表: ");
        System.out.println("12 节\t\t34 节\t\t56 节\t\t78 节");
        System.out.println("HTML\t\tJAVA\t\t 思想品德\t\t 体育");
    }
}
```

Teacher 类:

```
Public  class  Teacher  implements  Schedule{
    private String name;  //教师姓名
    private String specialty;  //教师专业

    public void showInfo(){
        System.out.println(specialty+"专业, "+ name + "老师");
    }
    public void  schInfo() {
        showInfo();
        System.out.println("周一课表: ");
        System.out.println("12 节\t\t34 节\t\t56 节\t\t78 节");
        System.out.println("JSP\t\t \t\toffice\t\t ");

    }
    public String getName() {
        return name;
    }
    public void setName(String name) {
        this.name = name;
    }
    public String getSpecialty() {
        return specialty;
    }
    public void setSpecialty(String specialty) {
        this.specialty = specialty;
    }
}
```

JiaoWu 类:

```
public class JiaoWu {
    public void print(Schedule schedule){
        System.out.println("-------------------------------------------");
        schedule.schInfo();
    }
}
```

测试类：

```
public static void main(String[] args) {
    JiaoWu jiaoWu = new JiaoWu();
    ComputerSch computerSch = new ComputerSch();
    computerSch.setName("计算机应用");
    computerSch.setQuantity(20);
    jiaoWu.print(computerSch);

    SoftwareSch softwareSch = new SoftwareSch();
    softwareSch.setName("软件");
    softwareSch.setQuantity(15);
    jiaoWu.print(softwareSch);

    Teacher teacher = new Teacher();
    teacher.setName("张三");
    teacher.setSpecialty("软件");
    jiaoWu.print(teacher);

}
```

执行效果：

```
Problems  @ Javadoc  Declaration  Console  Progress  LogCat
周一课表：
12节            34节            56节            78节
组装维护        SQLServer       JAVA            体育
------------------------------------------------------------
软件班，人数：15
周一课表：
12节            34节            56节            78节
HTML            JAVA            思想品德        体育
------------------------------------------------------------
软件专业，张三老师
周一课表：
12节            34节            56节            78节
JSP                             office
```

➢ 练一练：

扩展该程序，现在需要将每个班级期末的实训课表显示出来（每个班级的实训课表不一样），该如何修改程序？

提示：1．将实训课表定义为接口。

2．将显示实训课表方法放入 schInfo()中。

项目实训与练习

一、操作题

求三角形、圆形、正方形、梯形面积，要求设计一个公共的父类，有求面积的方法，通过继承产生各种形状的子类。

二、选择题

1．定义一个接口必须使用的关键字是（　　）。

A．public　　B．class　　C．interface　　D．static

2．接口是Java面向对象的实现机制之一，以下说法正确的是（　　）。

A．Java支持多重继承，一个类可以实现多个接口

B．Java只支持单重继承，一个类可以实现多个接口

C．Java只支持单重继承，一个类可以实现一个接口

D．Java支持多重继承，但一个类只可以实现一个接口

3．下列有关抽象类与接口的叙述中正确的是哪一个（　　）。

A．抽象类中必须有抽象方法，接口中也必须有抽象方法

B．抽象类中可以有非抽象方法，接口中也可以有非抽象方法

C．含有抽象方法的类必须是抽象类，接口中的方法必须是抽象方法

D．抽象类中的变量定义时必须初始化，而接口中不是

4．在Java中，下列说法正确的是（　　）。

A．一个子类可以有多个父类，一个父类也可以有多个子类

B．一个子类可以有多个父类，但一个父类只可以有一个子类

C．一个子类只可以有一个父类，但一个父类可以有多个子类

D．上述说法都不对

5．Father和Son是两个Java类，下列哪一选项正确地标识出Father是Son的父类（　　）。

A．class Son implements Father　　B．class Father implements Son

C．class Father extends Son　　D．class Son extends Father

三、填空题

1．接口中所有的属性均为________、________和________。

2．________方法是一种仅有方法声明，没有具体方法体和操作实现的方法，该方法必须在________类中定义。

3．在Java中，能实现多重继承效果的方式是________。

4．在Java程序中，通过类的定义只能实现________重继承，但通过________的定义可以实现多重继承关系。

强壮的计算器

项目目标

通过本任务的学习掌握异常处理的流程，学会异常捕获、异常处理、抛出异常的方法，能够利用异常处理机制处理程序中可能出现的异常。

项目内容

处理计算器程序可能出现的异常，不好处理的异常可以交给调用者处理，提高计算器程序的健壮性。

任务1 编写健壮的程序：异常处理

如果一个程序在使用过程中出现问题，导致程序崩溃，这时的用户体验是非常差的。而在编写程序时，最头疼的就是各种各样的错误，有时候编码不严谨。有些情况没有考虑到导致程序出错；有时候调用一些方法时，不了解这个方法的使用方式导致程序出错，这些错误是无处不在的，令人防不胜防。

✧ 需求分析

可以利用 java 提供的异常处理机制，来提高程序的健壮性，让用户满意。同时异常处理可以让程序在出错的情况下不会中途退出。现在用一个简单的计算器程序来理解异常处理机制。

下图是程序没有错误情况的执行效果：

```
Problems  @ Javadoc  Declaration  Console ☒  Progress
<terminated> Test (5) [Java Application] D:\Program Files (x86)\Java\j
请输入第一个数：5
请输入运算符(支持+、-、*、/)：+
请输入第一个数：3
计算结果为：8
```

下图是程序有错误情况的执行效果：

```
Problems  @ Javadoc  Declaration  Console ☒  Progress
<terminated> Test (5) [Java Application] D:\Program Files (x86)\Java\jre
请输入第一个数：a
异常：输入不为整数！
程序执行结束，谢谢使用！！
```

✧ 知识准备

1. 技能解析

处理异常主要的思路：首先在编写一段代码或调用别人写好的方法时，这段代码执行某些有风险的任务，可能会在运行期间出状况；必须认识到该方法是有风险的。这时需写出处理这些“意外状况”的代码。

Java 的异常捕获与处理是通过 5 个关键字来实现的：try、catch、finally、throw、throws。

关键字 try 构成的 try 语句块执行可能产生异常的代码。

关键字 catch 构成的 catch 语句块捕获异常，然后对异常进行所需的处理。

关键字 finally 构成的 finally 语句块完成一些资源释放、清理的工作，如关闭 try 程序块中所有打开的文件、断开网络连接。

在异常处理中，经常使用异常对象的方法。使用 getMessage()方法返回保存在某个异常中的描述字符串，使用 printStackTrace ()方法把调用堆栈的内容打印出来。

关键字 throw 用于手动抛出异常，throws 用于声明方法可能要抛出的各种异常，将在任务十四进行学习。

所以最主要的是能够预估该程序会出现哪些异常，从而对可能发生的异常做预先的处理。**异常处理的语句结构：**

```
try{
//try 语句块，可能产生异常的代码。
}catch(异常类型  异常引用变量){
//catch 语句块，处理异常的代码。捕获异常。
}finally{
        //finally 语句块，释放资源的代码。无论是否发生异常，代码都会执行。
}
```

在语句结构中，try 和 catch 部分是必需的，并且 catch 部分可以有多个，finally 语句块是可选项，可以没有。

异常处理的执行流程如图 3-1 所示。

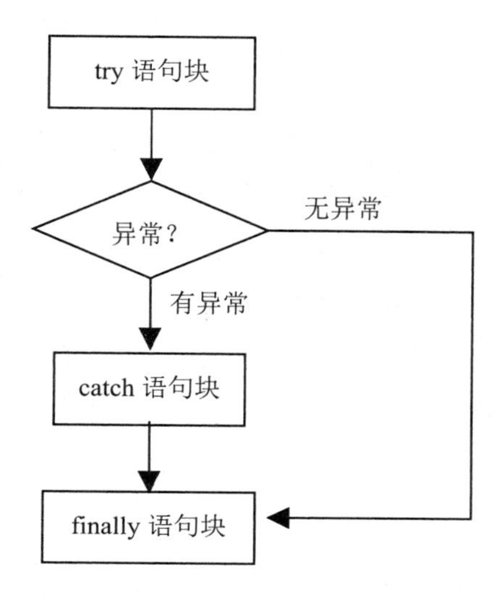

图 3-1　异常处理流程图

从异常处理的执行流程图中可以看出，当 try 语句块引发异常时，将会抛出异常对象，然后，在 catch 语句块中捕获异常对象，进行异常处理。如果无法捕获抛出的异常对象，则会发生错误，程序停止运行。

如果 try 语句块没有引发异常，catch 语句块将不执行。但是，**无论有没有异常抛出，finally 语句块总是被执行。**

2．知识解析

（1）异常处理机制　Java 使用异常处理机制为程序提供了异常处理的能力。所谓异常处理，就是在程序中预先想好对异常的处理办法，当程序运行出现异常时，对异常进行处理，处理完毕，程序继续运行。

Java 异常处理机制由捕获异常和处理异常两部分组成。当出现了异常事件，就会生成一个异常对象，传递给运行时系统，这个产生和提交异常的过程称为抛出（throw）异常。

在运行时系统得到异常对象，将会寻找处理异常的方法，把当前异常对象交给该方法处理，这一过程称为捕获（catch）异常。**捕获异常的过程中要求预先设定好捕获的类型，如果异常类型不匹配将不能成功捕获异常。**

如果没有找到可以捕获异常的方法，则运行时系统将终止，程序退出运行状态。

（2）异常的分类　Java 中，异常由类来表示，异常类的父类是 Throwable 类。Throwable 类有两个直接子类 Error 类和 Exception 类。Error 类表示程序运行时较少发生的内部系统错误，程序员无法处理。Exception 类表示程序运行时程序本身和环境产生的异常，可以捕获和处理。异常类继承结构如图 3-2 所示。

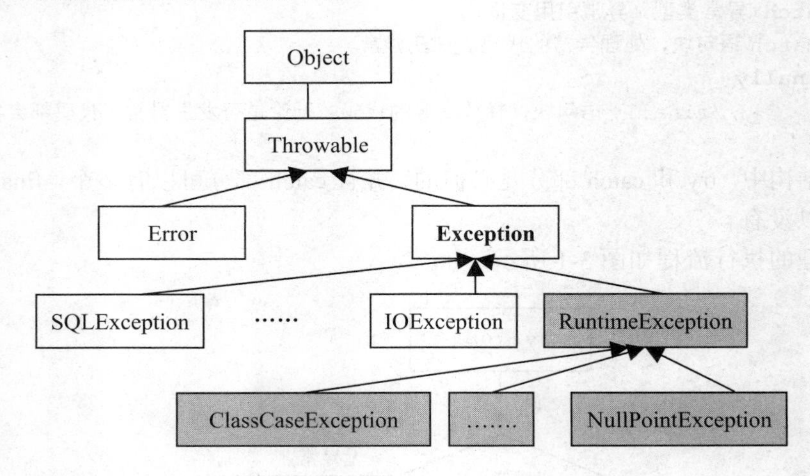

图 3-2　异常类继承结构

从图 3-2 中可以看到异常分为 RuntimeException 也叫运行时异常（已经用灰色标出），以及非运行时异常。运行时异常在编译器中是没有提示的，可以通过编译，这种异常可以通过改进代码实现来避免；非运行时异常无法通过编译，在编译器中会有提示。

✧ 编码实施

这是一个没有任何异常处理的简易计算器：

```
public static void main(String[] args) {
        Scanner input = new Scanner(System.in);
```

```
        System.out.print("请输入第一个数：");
        int a = input.nextInt();
        System.out.print("请输入运算符(支持+、-、*、/)：");
        String operator = input.next();
        System.out.print("请输入第一个数：");
        int b = input.nextInt();

        System.out.print("计算结果为：");
        if(operator.equals("+")){
            System.out.println("" + (a+b));
        }else if(operator.equals("-")){
            System.out.println("" + (a-b));
        }else if(operator.equals("*")){
            System.out.println("" + (a*b));
        }else if(operator.equals("/")){
            System.out.println("" + (a/b));
        }
    }
```

1．练一练

这个简易计算器是没有编译错误的，那么用户在使用过程中会出现哪些错误？

2．利用 try...catch...处理异常

首先可以简单粗暴地用 Exception 对象处理所有异常，Exception 对象是我们异常的父类，用 Exception 可以代表所有异常类型。

示例代码：

```
public static void main(String[] args) {
        Scanner input = new Scanner(System.in);
        try{        → try 语句块中的代码是我们需要执行的代码
            System.out.print("请输入第一个数：");
            int a = input.nextInt();
            System.out.print("请输入运算符(支持+、-、*、/)：");
            String operator = input.next();
            System.out.print("请输入第一个数：");
            int b = input.nextInt();
            System.out.print("计算结果为：");
            if(operator.equals("+")){
                System.out.println("" + (a+b));
            }else if(operator.equals("-")){
                System.out.println("" + (a-b));
            }else if(operator.equals("*")){
                System.out.println("" + (a*b));
            }else if(operator.equals("/")){
                System.out.println("" + (a/b));
            }
```

```
} catch(Exception e){
    System.err.println("程序运行出错！请联系管理员，错误代码：");
    e.printStackTrace();
  }
}
```

catch 语句块，catch 后面括号中是异常的类型

当程序有异常出现时，会执行 catch 语句块中的内容

由于 input.nextInt();只能接收整型数据，所以当我们输入"a"时会出现异常，运行效果：

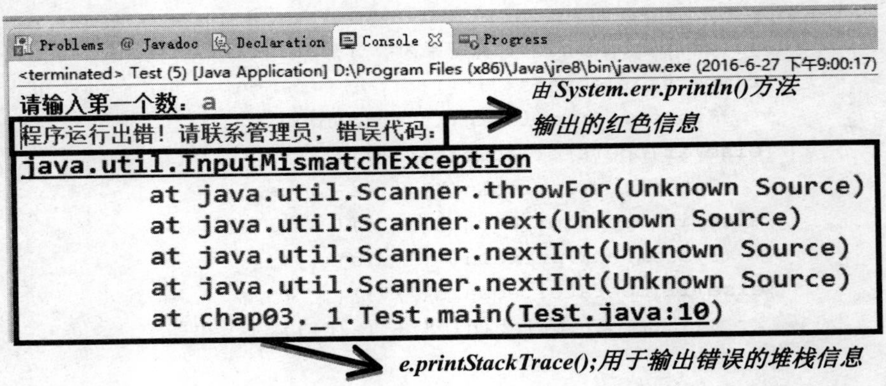

```
Problems @ Javadoc Declaration Console Progress
<terminated> Test (5) [Java Application] D:\Program Files (x86)\Java\jre8\bin\javaw.exe (2016-6-27 下午9:00:17)
请输入第一个数：a
程序运行出错！请联系管理员，错误代码：
java.util.InputMismatchException
        at java.util.Scanner.throwFor(Unknown Source)
        at java.util.Scanner.next(Unknown Source)
        at java.util.Scanner.nextInt(Unknown Source)
        at java.util.Scanner.nextInt(Unknown Source)
        at chap03._1.Test.main(Test.java:10)
```

由 System.err.println()方法输出的红色信息

e.printStackTrace();用于输出错误的堆栈信息

当程序出现异常时，同时输出文字信息和错误堆栈信息是很有必要的，文字信息可以给用户更好的使用体验，而错误堆栈信息能够让技术人员更好地处理错误。

3．finally 语句块

在异常处理中还有一个 finally 块，它是无论如何都会执行的部分。也就是说 finally 块是用来存放不管有没有异常都会执行的部分。在一个程序中有时有些代码是必须要执行的，就好像在家里做饭不管饭菜好不好吃都要把火关了一样。

我们编写的计算器程序执行结束时没有提示，为了提高用户体验，现在加入此项功能。

示例代码：

```
public static void main(String[] args) {
        Scanner input = new Scanner(System.in);
        try{
            System.out.print("请输入第一个数：");
            int a = input.nextInt();
            System.out.print("请输入运算符(支持+、-、*、/)：");
            String operator = input.next();
            System.out.print("请输入第一个数：");
            int b = input.nextInt();
            System.out.print("计算结果为：");
            if(operator.equals("+")){
                System.out.println("" + (a+b));
            }else if(operator.equals("-")){
                System.out.println("" + (a-b));
            }else if(operator.equals("*")){
                System.out.println("" + (a*b));
            }else if(operator.equals("/")){
                System.out.println("" + (a/b));
            }
        }catch(Exception e){
```

```
            System.err.println("程序运行出错！请联系管理员，错误代码：");
            e.printStackTrace();
    }finally{
            System.out.println("程序执行结束，谢谢使用!! ");
        }
}
```

finally 语句块，存放的是不管是否异常都会执行的代码

程序正常运行时执行效果：

```
Problems  @ Javadoc  Declaration  Console  Progress
<terminated> Test (5) [Java Application] D:\Program Files (x86)\Java\jre
请输入第一个数：3
请输入运算符(支持+、-、*、/)：+
请输入第一个数：5
计算结果为：8
程序执行结束，谢谢使用！！
```

程序出现异常时执行效果：

```
Problems  @ Javadoc  Declaration  Console  Progress
<terminated> Test (5) [Java Application] D:\Program Files (x86)\Java\jre8\bin\javaw.exe (20
请输入第一个数：3
请输入运算符(支持+、-、*、/)：/
请输入第一个数：0
计算结果为：程序运行出错！请联系管理员，错误代码：
java.lang.ArithmeticException: / by zero
        at chap03._1.Test.main(Test.java:23)
程序执行结束，谢谢使用！！
```

✧ 调试运行

1．在 try 块中，某处代码出现异常，则这句代码下面的语句将不会执行。

示例代码：

```
public static void main(String[] args) {
        Scanner input = new Scanner(System.in);
    try{
            ......
代码略
......
            System.out.println("程序执行结束，谢谢使用!! ");
        }catch(Exception e){
            System.err.println("程序运行出错！请联系管理员，错误代码：");
            e.printStackTrace();
        }
}
```

前面的代码不变，将 finally 块中的语句放到 try 块的最后，当程序出现异常时，这句话将无法执行

执行结果：

```
Problems @ Javadoc Declaration Console Progress
<terminated> Test (5) [Java Application] D:\Program Files (x86)\Java\jre8\bin\javaw.exe (2016-6-27 下午10:54:51)
请输入第一个数：a
程序运行出错！请联系管理员，错误代码：
java.util.InputMismatchException
	at java.util.Scanner.throwFor(Unknown Source)
	at java.util.Scanner.next(Unknown Source)
	at java.util.Scanner.nextInt(Unknown Source)
	at java.util.Scanner.nextInt(Unknown Source)
	at chap03._1.Test.main(Test.java:10)
```

2．异常分为很多种类型，如果异常类型不匹配将无法捕获该异常。

示例代码：

```
public static void main(String[] args) {
    Scanner input = new Scanner(System.in);
        try{
            ......
        代码略
            ......
    }catch(ArithmeticException  e ){
            System.err.println("程序运行出错！请联系管理员，错误代码：");
            e.printStackTrace();
        }finally{
            System.out.println("程序执行结束，谢谢使用！！");
        }
}
```

将异常类型修改为ArithmeticException，它是Exception的子类，用于处理算数异常，比如除数为零

这时如果输入一个“a”，会得到下面的效果：

```
Problems @ Javadoc Declaration Console Progress
<terminated> Test (5) [Java Application] D:\Program Files (x86)\Java\jre8\bin\javaw.exe (2016-6-27 下午10:57:38)
请输入第一个数：a
程序执行结束，谢谢使用！！
Exception in thread "main" java.util.InputMismatchException
	at java.util.Scanner.throwFor(Unknown Source)
	at java.util.Scanner.next(Unknown Source)
	at java.util.Scanner.nextInt(Unknown Source)
	at java.util.Scanner.nextInt(Unknown Source)
	at chap03._1.Test.main(Test.java:10)
```

出现异常时的文字提示并没有打印出来，这说明 catch 块中的语句并没有执行。

常见的异常类型见表 3-1 和表 3-2。

表 3-1　非运行时异常

异常类型	异常层次结构的根类
ClassNotFoundException	找不到类或接口产生的异常
CloneNotSupportedException	使用对象的 clone()方法，但无法执行 Cloneable 产生的异常
IllegalAccessException	类定义不明确产生的异常
InstantiationException	使用 newInstance()方法试图建立一个类 instance 时产生的异常
InterruptedException	目前线程等待执行，另一线程中断目前线程产生的异常
NoSuchMethodException	找不到方法
SecurityException	违反安全产生的异常。当 applet 企图执行由于浏览器的安全设置而不允许的动作时产生
FileNotFoundException	找不到文件
EOFException	文件结束
SQLException	数据库访问的异常

表 3-2 运行时异常

异常	子类	说明
RuntimeException		运行异常，在 JVM 正常运行时抛出异常的父类
ArithmeticException		算术异常。算术运算产生的异常，如零作除数
ArrayStoreException		存入数组的内容与数组类型不一致时产生的异常
ClassCastException		类对象强制转换造成不当类对象产生的异常。如，类 C 对象 c 强制成类 A，而 c 既不是 A 的实例，也不是 A 的子类的实例
IllegalArgumentException（方法接收到非法参数，父类）	IllegalThreadStateException	线程在不合理状态下运行产生的异常
	NumberFormatException	字符串转换为数值产生的异常，如“8”正常，“s”异常
IllegalMonitorStateException		线程等候或通知对象时产生的异常
IndexOutOfBoundsException（索引越界产生的异常，父类）	ArrayIndexOutOfBoundsException	数组索引越界产生的异常
	StringIndexOutOfBoundsException	企图访问字符串中不存在的字符位置时产生
NegativeArraySizeException		创建数组时长度为负数
NullPointerException		空指针异常。企图引用值为 null 的对象时产生的异常

✧ 维护升级

在计算器程序中，异常处理方法非常简单，并不能给用户详细的错误提示，使用户不知道什么地方出现问题。现在让我们可以更加细致地处理这些异常：

示例代码：

```
public static void main(String[] args) {
        Scanner input = new Scanner(System.in);
        try{
            System.out.print("请输入第一个数: ");
            int a = input.nextInt();
            System.out.print("请输入运算符(支持+、-、*、/): ");
            String operator = input.next();
            System.out.print("请输入第一个数: ");
            int b = input.nextInt();
            System.out.print("计算结果为: ");
            if(operator.equals("+")){
                System.out.println("" + (a+b));
            }else if(operator.equals("-")){
                System.out.println("" + (a-b));
            }else if(operator.equals("*")){
                System.out.println("" + (a*b));
            }else if(operator.equals("/")){
                System.out.println("" + (a/b));
            }else{
                System.out.println("运算符输入不正确!~");
        }
    }catch (InputMismatchException ie) {
        System.out.println("异常:输入不为整数!");
    }catch (ArithmeticException ae) {
            System.out.println("异常:除数不能为零!");
```

为 input.nextInt() 传入的数据不为数字时产生的异常

当 a/b 进行计算时，b 为 0 会产生的异常

```
    }catch (Exception e) {
        System.out.println("其他异常请通知管理员:"+e.getMessage());
    }finally{
        System.out.println("程序执行结束，谢谢使用!! ");
    }
}
```

最后用异常的父类处理其他没有考虑到的异常。

注意：catch 捕获异常是按照顺序捕获，比如上例中先捕获 InputMismatchException 的异常，接着捕获 ArithmeticException，最后捕获 Exception；所以捕获 Exception 类型的异常必须放在最后。

现在当用户输入的数据出现错误时，程序会给出出错的原因：

```
Problems @ Javadoc Declaration Console Progress
<terminated> Test (5) [Java Application] D:\Program Files (x86)\Java\jre
请输入第一个数：a
异常:输入不为整数!
程序执行结束，谢谢使用! !
```

任务 2　别人的异常：抛出异常

任务一中我们知道了产生异常时该如何处理；还有一些情况下我们调用的方法本身也会产生异常，比如在后面会学习到的 JDBC 中，会有一些方法调用的时候就会有异常，这时必须处理它们。这是编程人员手动**抛出的异常**。

✧ 需求分析

我们花在异常处理上的时间比创建和抛出异常的时间要多很多，而且有些情况下异常并不适合在编码的时候解决。让方法调用者去处理异常也是一种常用的处理方法。如何才能让方法调用者去处理异常呢？

✧ 知识准备

要用到声明异常。声明异常的格式如下：

```
[修饰符]<返回类型>  方法名([参数列表])  throws  异常列表{
    [throw  异常对象];   //可以根据实际情况选择是否创建异常
}
```

其中 **throws** 是关键字。

如果当前 java 提供的异常不符合实际编程需要，那么我们可以自己创建一个新型异常。如果想要自定义异常只需要继承异常类型即可。自定义异常的格式为：

```
class  自定义异常名   extends  Exeption{
        //有参构造方法;
        //无参构造方法;
}
```

✧ 编码实施

现在我们不再自己处理异常，而是将异常抛出，让调用者去处理它。

示例代码：

将异常抛出

```
public void calc() throws ArithmeticException, InputMismatchException {
        Scanner input = new Scanner(System.in);
        int a;
        int b;
        String operator;
        try{
            System.out.print("请输入第一个数：");
            a = input.nextInt();
            System.out.print("请输入运算符(支持+、-、*、/)：");
            operator = input.next();
            System.out.print("请输入第一个数：");
            b = input.nextInt();
            System.out.print("计算结果为：");
            if(operator.equals("+")){
                System.out.println("" + (a+b));
            }else if(operator.equals("-")){
                System.out.println("" + (a-b));
            }else if(operator.equals("*")){
                System.out.println("" + (a*b));
        }else if(operator.equals("/")){
            if(b==0){
                throw new ArithmeticException("除数不能为 0");
                }else{
                    System.out.println("" + (a/b));
                }
            }else{
            System.out.println("运算符输入不正确!~");
        }
        }catch(Exception e){
        throw new InputMismatchException("输入不为整数");
        }
}
```

当 b 等于 0 时，创建异常对象，并抛出

捕获异常后并没有处理，而是创建异常对象后抛出

测试类：

```
public static void main(String[] args) {
        Calculator calculator = new Calculator();
        try{
            calculator.calc();
        }catch (InputMismatchException ie) {
            ie.printStackTrace();
        }catch (ArithmeticException ae) {
            ae.printStackTrace();
        }

}
```

执行效果：

```
Problems  @ Javadoc  Declaration  Console ⊠  Progress  LogCat
<terminated> Test (3) [Java Application] D:\Program Files (x86)\Java\jre8\bin\javaw.exe (2016-6-28 上午10:
请输入第一个数：a
java.util.InputMismatchException: 输入不为整数
        at chap03._1.Calculator.calc(Calculator.java:39)
        at chap03._1.Test.main(Test.java:11)
```

✧ 调试运行

RuntimeException，也叫运行时异常，不会影响编译，可以不做处理。

非运行时异常，不能通过编译，只能处理或者抛出。

运行时异常示例代码：

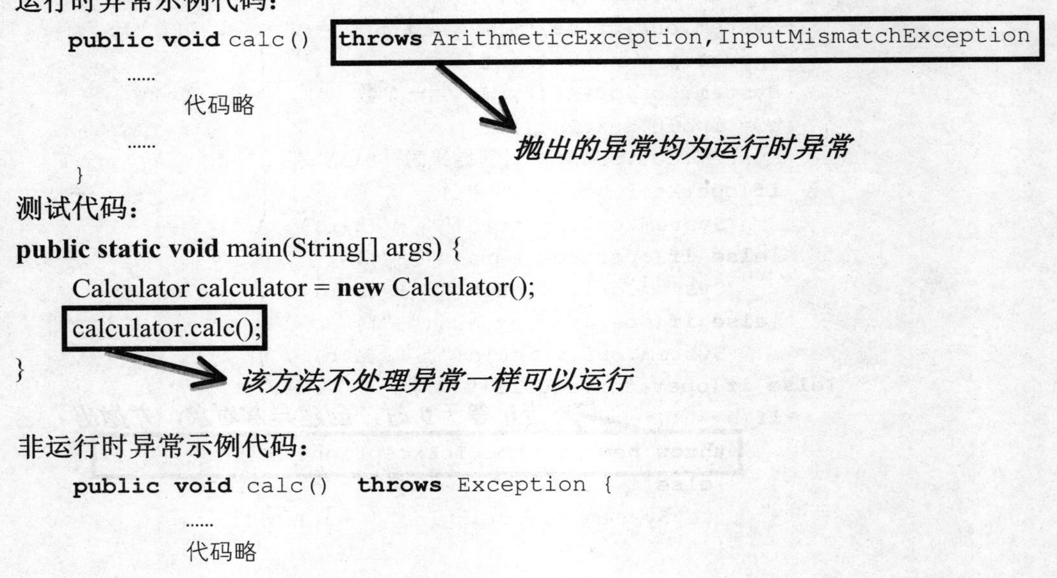

```
public void calc() throws ArithmeticException,InputMismatchException {
    ......
        代码略
    ......
}
```

抛出的异常均为运行时异常

测试代码：

```
public static void main(String[] args) {
    Calculator calculator = new Calculator();
    calculator.calc();
}
```

该方法不处理异常一样可以运行

非运行时异常示例代码：

```
public void calc()  throws Exception {
        ......
        代码略
        ......
}
```

测试代码：

```
public class Test {
    public static void main(String[] args) {
        Calculator calculator = new Calculator();
        calculator.calc();
    }
}
```

```
Unhandled exception type Exception
2 quick fixes available:
  Add throws declaration
  Surround with try/catch
```

✧ 维护升级

Java 提供的异常类型无法做到精细的提示，有时并不能满足需求，这时还可以用自定义异常来实现更加细致的错误提示信息。

示例代码：

Calculator1 类：

```
public class Calculator1 {
public void calc() throws MyException{
        Scanner input = new Scanner(System.in);
        int oper1 ;
        int oper2 ;
        String operator;
        String a;
        String b;
    System.out.print("请输入第一个整数: ");
        a = input.next();
    try{
        oper1 = Integer.valueOf(a).intValue();
    }catch (NumberFormatException e){
            throw new MyException("输入错误,第一个数必须是整数!",a );
        }
    System.out.print("请输入运算符(支持+、-、*、/): ");
    operator = input.next();
        System.out.print("请输入第二个整数: ");
        b = input.next();
        try{
            oper2 = Integer.valueOf(b).intValue();
        }catch (NumberFormatException e){
            throw new MyException("输入错误,第二个数必须是整数!", b );
        }
        System.out.print("计算结果为: ");
        if(operator.equals("+")){
            System.out.println("" + (oper1+oper2));
        }else if(operator.equals("-")){
            System.out.println("" + (oper1-oper2));
        }else if(operator.equals("*")){
            System.out.println("" + (oper1*oper2));
        }else if(operator.equals("/")){
            if(oper2==0){
                throw new MyException("除数不能为0",String.valueOf(oper2));
            }else{
                System.out.println("" + (oper1/oper2));
            }
        }else{
            throw new MyException("运算符输入不正确!~");
        }
}
}
```

将 String 类型转换成整数，如果无法转换则会抛出异常，说明用户输入出错

当转换出错时会抛出我们自定义的异常，并且传递两个参数，第一个参数为提示信息，第二个参数为用户输入的数据

MyException 类：

```
public class MyException extends Exception{
    private String oper;
    private String message;

    public MyException(){
```

```
        super();
    }
    public MyException(String str){
        super(str);
    }
    public MyException(String str, String oper){
        this.message = str;
        this.oper = oper;
    }
    public String getMessage(){
        return message+", 你输入的数是:" + oper;
    }
}
```

构造方法，用于接受错误提示信息和用户输入的数据

测试类：

```
public static void main(String[] args) {
        Calculator1 calculator1 = new Calculator1();
        try {
            calculator1.calc();
        } catch (MyException e) {
            System.out.println( e.getMessage() );
        }
}
```

捕获异常，并显示错误信息

执行结果：

项目实训与练习

一、操作题

设计一个自定义异常，表示 String 对象的长度太长，让用户输入字符串，如果超过 20 字符，就抛出这个异常。

二、选择题

1．finally 语句块中的代码（　　）。

A．总是被执行

B．当 try 语句块后面没有 catch 时，finally 中的代码才会执行

C．异常发生时才执行

D．异常没有发生时才被执行

2．自定义异常类时，可以继承的类是（　　）。

A．Error　　B．Applet

C．Exception 及其子类　　D．AssertionError

3．对于 try{,,,,}catch 子句的排列方式，下列正确的一项是（　　）。

A．子类异常在前，父类异常在后

B．父类异常在前，子类异常在后

C．只能有子类异常

D．父类异常与子类异常不能同时出现

4．使用 catch(Exception e)的好处是（　　）。

A．只会捕获个别类型的异常

B．捕获 try 语句块中产生的所有类型的异常

C．忽略一些异常

D．执行一些程序

三、填空题

1．一个 try 语句块后必须跟________语句块，________语句块可以没有。

2．自定义异常类必须继承________类及其子类。

3．异常处理机制允许根据具体的情况选择在何处处理异常，可以在________捕获并处理，也可以用 throws 子句把它交给________处理。

项目 4 复杂的数据

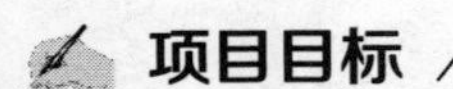

项目目标

重点掌握 ArrayList、HashMap 两种集合的使用场景，学会使用迭代器遍历集合，掌握泛型的基本使用。

项目内容

编写歌曲管理程序，实现歌曲的添加、删除以及显示选中歌曲的详细信息。

任务 1　歌曲管理程序：ArrayList、HashMap 集合

集合是非常重要的部分，它类似于数组，可以成批量地操作数据。随着编程学习的越来越深入，集合使用的会越来越多，利用集合可以非常方便地添加元素、删除特定元素、搜索元素、排序、去除重复项等。

✧ 需求分析

制作一个歌单管理程序，可以显示现有歌单，同时能够对歌曲进行插入、删除、排序、修改操作。数组的长度是固定的，想要给数组添加元素，或者删除元素可不容易，但是用集合完成要容易得多。

执行效果：

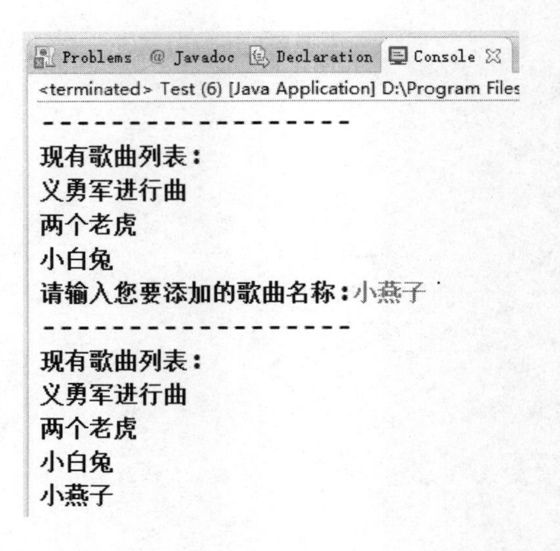

```
Problems  @ Javadoc  Declaration  Console
<terminated> Test (6) [Java Application] D:\Program File
------------------
现有歌曲列表：
1.义勇军进行曲
2.两个老虎
3.小白兔
请输入你要删除歌曲的序号：3
------------------
现有歌曲列表：
1.义勇军进行曲
2.两个老虎
```

✧ 知识准备

1. 技能解析

（1）ArrayList 是最常用的集合类，它的存储结构和数组类似，但是它的长度可以动态增长，也可以将它称为**动态数组**。Array list 常用方法见表 4-1。

表 4-1　ArrayList 常用方法

返回值类型	方法名及作用
boolean	**add**(E　e) 将指定的元素添加到此列表的尾部
void	**clear**() 移除此列表中的所有元素
int	**indexOf**(Object　o) 返回此列表中首次出现的指定元素的索引，或如果此列表不包含元素，则返回-1
boolean	**isEmpty**() 如果此列表中没有元素，则返回 true
int	**lastIndexOf**(Object　o) 返回此列表中最后一次出现的指定元素的索引，或如果此列表不包含索引，则返回-1
E	**remove**(int　index) 移除此列表中指定位置上的元素
boolean	**remove**(Object　o) 移除此列表中首次出现的指定元素（如果存在）
E	**set**(int index, Eelement) 用指定的元素替代此列表中指定位置上的元素
int	**size**() 返回此列表中的元素数
Object[]	**toArray**() 按适当顺序（从第一个到最后一个元素）返回包含此列表中所有元素的数组

（2）LinkedList 链表集合，针对经常插入和删除中间元素所设计的高效率集合。LinkedList 提供额外的 get、remove、insert 方法在 LinkedList 的首部或尾部。

（3）HashMap 映射集合，可用成对的 key/value 来保存与退出。

（4）HashSet，防止重复的集合，可以快速寻找相符的元素。

2．知识解析

（1）实现 Set 和 List 接口的常用集合 见图 4-1。

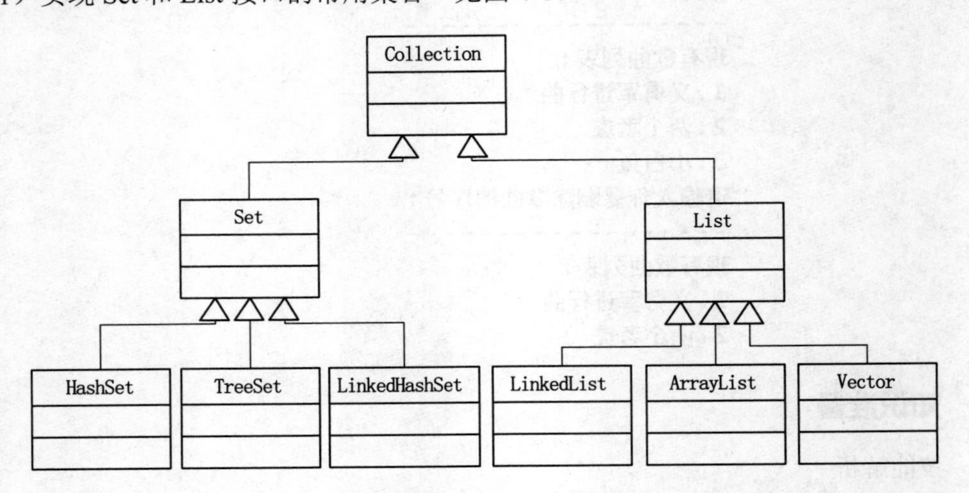

图 4-1 实现 Set 和 List 接口的常用集合

（2）实现 Map 接口的常用集合 见图 4-2。

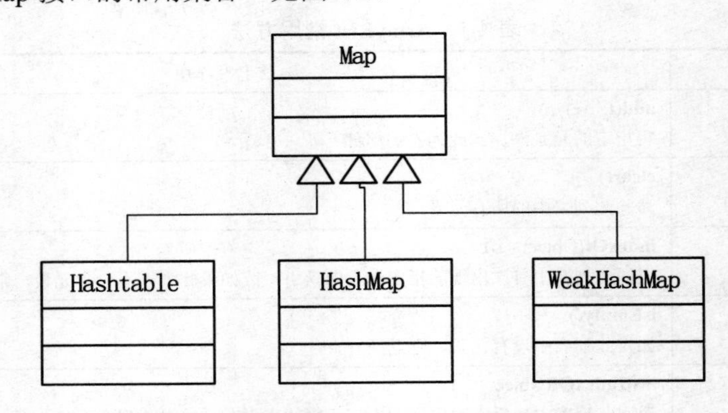

图 4-2 实现 Map 接口的常用集合

（3）不同集合的作用

List 接口下的集合存储数据是顺序存储。可以通过索引随机访问元素。

Set 接口下的集合元素不会重复。

Map 接口下的集合均是通过键值对的方式存储数据的。

集合件特性对比见表 4-2。

表 4-2 集合件特性对比

接口	简述	实现	操作特性	元素要求
Set	成员不能重复	HashSet	无序地遍历成员	元素可为任意 Object 子类的对象
		TreeSet	有序地遍历成员	成员要求实现 caparable 接口，或者使用 Comparator 构造 TreeSet。成员一般为同一类型
		LinkedHashSet	外部按成员的插入顺序遍历成员	成员与 HashSet 成员类似

续表

接口	简述	实现	操作特性	元素要求
List	通过索引对成员的随机访问	ArrayList	提供快速的基于索引的成员访问，对尾部成员的增加和删除支持较好	成员可为任意 Object 子类的对象
		LinkedList	对列表中任何位置的成员的增加和删除支持较好，但对基于索引的成员访问支持性能较差	成员可为任意 Object 子类的对象
Map	保存键值对成员，基于键找值操作	HashMap	能满足用户对 Map 的通用需求	键成员可为任意 Object 子类的对象
		TreeMap	支持对键有序地遍历	键成员要求实现 caparable 接口，或者使用 Comparator 构造 TreeMap。键成员一般为同一类型
		LinkedHashMap	保留键的插入顺序	成员可为任意 Object 子类的对象，但如果覆盖了 equals 方法，同时注意修改 hashCode 方法
		WeakHashMap	其行为依赖于垃圾回收线程，没有绝对理由则少用	

✧ 编码实施

（1）简易版歌曲管理程序，仅实现了歌曲的显示功能：

SongManager 类：

```
public class SongManager {
    final static String [] songs = {"义勇军进行曲","两个老虎","小白兔"};
    List songList = new ArrayList();

    public SongManager(){
        for(int i=0 ; i<songs.length ; i++){
            songList.add(songs[i]); //遍历数组，将数组中的元素添加到集合中
        }
    }

    public void show(){
System.out.println("------------------");
        System.out.println("现有歌曲列表:");
        for(int i=0 ; i<songList.size() ; i++){
            String song = (String) songList.get(i);//从集合中取出元素
            System.out.println(song);
        }
    }
}
```

测试类：

```
public static void main(String[] args) {
        SongManager songManager = new SongManager();
        songManager.show();
}
```

执行效果：

```
Problems  @ Javadoc  Declaration  Console
-------------------
现有歌曲列表：
义勇军进行曲
两个老虎
小白兔
```

（2）加入添加歌曲功能。

在 **SongManager** 类中加入方法：

```
public void add(String song){
        songList.add(song);
}
```

测试类：

```
public static void main(String[] args) {
        Scanner input = new Scanner(System.in);
        SongManager songManager = new SongManager();
        songManager.show();

        System.out.print("请输入您要添加的歌曲名称:");
        String song = input.next();//输入歌曲名称
        songManager.add(song); //将歌曲添加到集合中
        songManager.show();
}
```

执行结果：

```
Problems  @ Javadoc  Declaration  Console
<terminated> Test (6) [Java Application] D:\Program Files
-------------------
现有歌曲列表：
义勇军进行曲
两个老虎
小白兔
请输入您要添加的歌曲名称：小燕子
-------------------
现有歌曲列表：
义勇军进行曲
两个老虎
小白兔
小燕子
```

练一练：

为歌曲管理系统加入删除功能。
提示：可以利用 remove(int index)方法删除指定元素

✧ 调试运行

（1）在使用集合时同样要注意索引越界问题：

如 SongManager 类中的 show()方法：

i<=songList.size()时，将会出现异常

```
public void show(){
    System.out.println("------------------");
        System.out.println("现有歌曲列表:");
    for(int i=0 ; i<=songList.size() ; i++){
            String song = (String) songList.get(i);//从集合中取出元素
            System.out.println(song);
        }
}
```

执行结果：

```
Problems  @ Javadoc  Declaration  Console  Progress
------------------
现有歌曲列表:
义勇军进行曲
两个老虎
小白兔
Exception in thread "main" java.lang.IndexOutOfBoundsException: Index: 3, Size: 3
        at java.util.ArrayList.rangeCheck(Unknown Source)
        at java.util.ArrayList.get(Unknown Source)
        at chap04._1.SongManager1.show(SongManager1.java:24)
        at chap04._1.Test.main(Test.java:9)
```

（2）目前从集合中获取数据需要进行强制类型转换操作(后续课程中会解决这个问题)。

SongManager 类中的 show()方法：

get()方法的返回值是 Object 类型，需要强制转换成需要的类型

```
public void show(){
        System.out.println("------------------");
    System.out.println("现有歌曲列表:");
        for(int i=0 ; i<=songList.size() ; i++){
        String song = (String) songList.get(i);  //从集合中取出元素
            System.out.println(song);
        }
}
```

✧ 维护升级

1．利用 Arrays.asList( )方法可以非常方便地将数组转换成集合。

示例代码：

```
public SongManager1(){
        for(int i=0 ; i<songs.length ; i++){
            songList.add(songs[i]); //遍历数组，将数组中的元素添加到集合中
        }
}
```

修改为：

```
public SongManager1(){
        songList = Arrays.asList(songs);
}
```

2. 实现删除功能可以利用 HashMap 集合来实现，也是非常的方便快捷。

示例代码：

```
public class SongManager2 {
    final static String [] songs = {"义勇军进行曲","两个老虎","小白兔"};
    Map songMap = new HashMap();
    public SongManager2(){
        songMap.put("1", songs[0]);//利用"键/值"的形式存储数据
        songMap.put("2", songs[1]);//第一个参数是键，第二个参数是值
        songMap.put("3", songs[2]);
    }
    public void show(){
        System.out.println("------------------");
        System.out.println("现有歌曲列表:");
        for(int i=1 ; i<=songMap.size() ; i++){
            String song = (String) songMap.get(""+i);//通过键从集合中取出值
            System.out.println(i+"." + song);
        }
    }
    public void delete(String id){
        songMap.remove(id);
    }
}
```

测试类：

```
public static void main(String[] args) {
    Scanner input = new Scanner(System.in);
    SongManager2 songManager = new SongManager2();
    songManager.show();
    System.out.print("请输入你要删除歌曲的序号:");
    String id = input.next();
    songManager.delete(id);
    songManager.show();
}
```

运行效果：

```
Problems  @ Javadoc  Declaration  Console
<terminated> Test (6) [Java Application] D:\Program File
------------------
现有歌曲列表:
1.义勇军进行曲
2.两个老虎
3.小白兔
请输入你要删除歌曲的序号:3
------------------
现有歌曲列表:
1.义勇军进行曲
2.两个老虎
```

上例中看起来没有问题，但是如果删除前两首歌曲时就会出现问题，如图 4-3 所示。

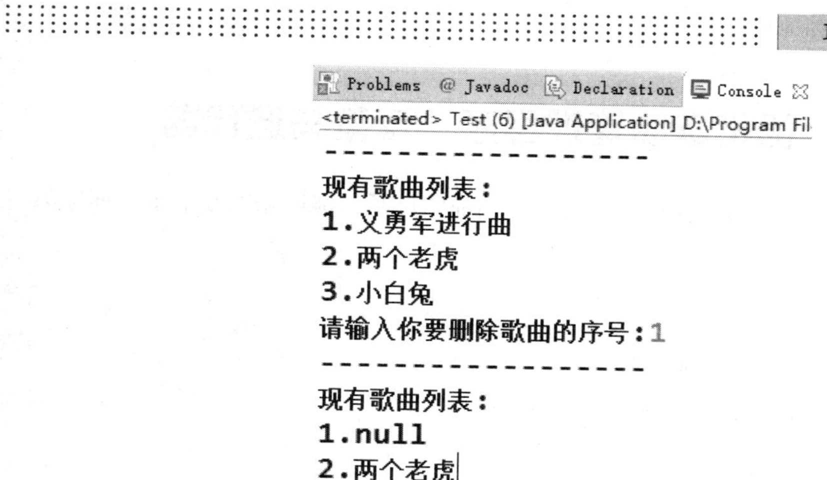

图 4-3 删除 1 号歌曲结果图

实际上已经将第一首歌曲删除了，集合长度变为 2，循环执行两次，显示了两首歌，但是却将为空的第一首歌显示出来了。这是由于显示的时候错误的拿到了第一首歌的“键”。示例代码：

```
public void show(){
        System.out.println("------------------");
        System.out.println("现有歌曲列表:");
    for(int i=1 ; i<=songMap.size() ; i++){
        String song = (String) songMap.get(""+i);      //通过键从集合中取出值
        System.out.println(i+"." + song);
    }
}
```

当删除一首歌时，集合长度变为 2，但是 get() 方法中的键还是从 1 开始，就可能导致为空的歌曲被显示。

要从根本上解决这个问题需要用到迭代器，会在下一个任务中学到，现在先暂时用另一种方式将歌曲显示出来。

将 SongManager2 类中的 show()方法改为：

```
public void show(){
        System.out.println("------------------");
        System.out.println("现有歌曲列表:");
        System.out.println(songMap);      //直接输入集合中的元素
}
```

执行效果：

```
Problems @ Javadoc Declaration Console Progress
<terminated> Test (6) [Java Application] D:\Program Files (x86)\Java\
------------------
现有歌曲列表：
{1=义勇军进行曲， 2=两个老虎， 3=小白兔}
请输入你要删除歌曲的序号：1
------------------
现有歌曲列表：
{2=两个老虎， 3=小白兔}
```

可以看到虽然显示效果不太理想，但是程序的功能实现了。

任务 2　优化歌曲管理程序：泛型与迭代器

本任务要用到泛型和迭代器，在程序开发中应用非常广泛，能够帮助我们更方便地编写程序。

✧ 需求分析

1．任务 1 的程序有两个小问题：

- 从集合获取数据时，必须强制类型转换。
- 删除歌曲时，显示效果无法自己调整。

2．使程序能够显示歌曲详细信息。

执行效果：

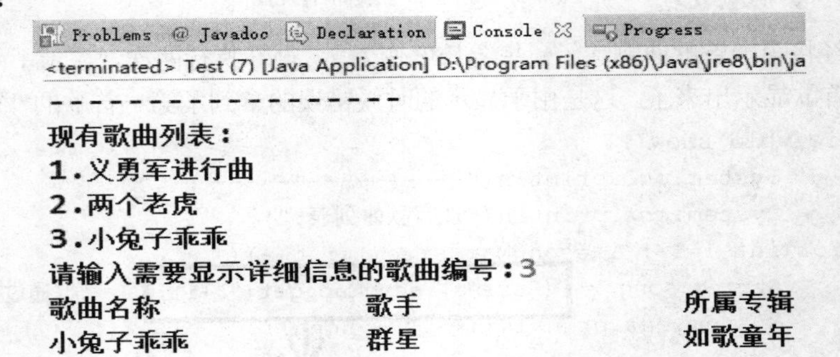

✧ 知识准备

1．技能解析

（1）Iterator 接口（迭代器）中常用方法。

- boolean　hasNext()：检查集合中是否还有元素，如果有则返回 true。
- Object　next()：将光标向后移动一个位置，并返回元素的值。
- void　remove()：删除集合中的当前元素。

Iterater 迭代器使用 next()方法来遍历集合，你可以想象集合中有一个游标，游标指在某一个元素上，next()方法每执行一次，游标都会向后移动。并且取出这个元素的值。如图 4-4 所示。

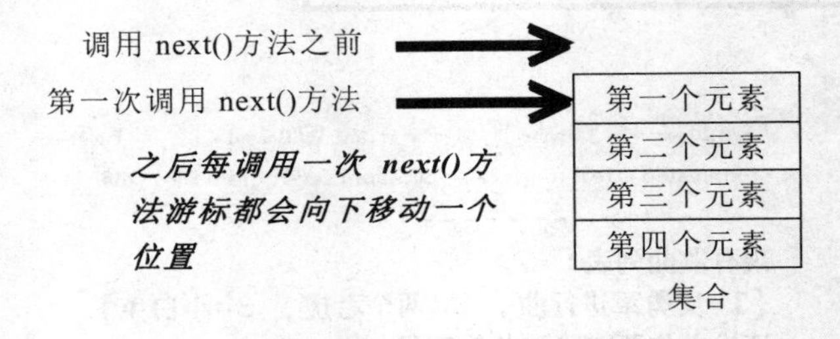

图 4-4　集合遍历过程图

（2）泛型。在定义集合时，可以使用泛型限定集合的数据类型。

如：

```
ArryList<参数化类型>  list = new ArrayList<参数化类型>();
```

2. 知识解析

（1）Iterator 接口（迭代器）。使用集合时，对集合的遍历几乎是一个必不可少的操作。Iterator 接口就专为遍历而生。几乎所有的集合类都实现了 Iterator，提供 Iterator 相关的 API，通过 Iterator 可以非常方便地的遍历集合。

（2）泛型。“参数化类型”即泛型，可以通过类似参数的方式指定数据类型。虽然集合能够存储任何类型的数据，但是从集合取出的数据都是 object 类型，导致原类型“消失”。而取出元素时，总进行强制类型转换就比较容易出错。而泛型就是为解决这个问题而生。

✧ 编码实施

1．解决任务一中显示歌曲问题。

示例代码：

```
public void show(){
        System.out.println("------------------");
        System.out.println("现有歌曲列表:");
        Set keySet = songMap.keySet(); //获取 map 集合中的所有 key,
//并且存放在一个 set 集合中
        Iterator iterator = keySet.iterator();      //得到 set 集合的迭代器
        while(iterator.hasNext()){      //判断下一个元素是否为空
            String key = (String) iterator.next(); //移动到下一个元素
                                                   //并获取其中存放的 key
            String song = (String) songMap.get(key);    //利用获取到的 key
//取出 map 集合中的元素
            System.out.println(key + "." + song);//将 key 和 value 显示出来
        }
}
```

测试类：

```
public static void main(String[] args) {
        Scanner input = new Scanner(System.in);
        SongManager3 songManager = new SongManager3();
        songManager.show();
        System.out.print("请输入你要删除歌曲的序号:");
        String id = input.next();
        songManager.delete(id);
        songManager.show();

}
```

2．上例中划框位置需要强制类型转换，这导致集合在使用时有些不方便，有时还会由于强制类型转换而出错，可以利用泛型解决这个问题。

示例代码：

使用泛型定义集合，就是在类名后面加入<限定数据类型1，限定数据类型2>，这样集合中只能存储限定数据类型的数据了

SongManager3 类：

```
public class SongManager3 {
    final static String [] songs = {"义勇军进行曲","两个老虎","小白兔"};
    Map<String,String> songMap = new HashMap<String,String>();
```

```
    public SongManager3(){
        songMap.put("1", songs[0]); //利用“键/值”的形式存储数据
        songMap.put("2", songs[1]); //第一个参数是键，第二个参数是值
        songMap.put("3", songs[2]);
    }
    public void show(){
        System.out.println("-------------------");
        System.out.println("现有歌曲列表:");
        Set<String>  keySet = songMap.keySet(); //获取map集合中的所有key，
                                                 并且存放在一个 set 集合中
        Iterator<String> iterator = keySet.iterator(); //得到 set 集合的
                                                        迭代器
        while(iterator.hasNext()){ //判断下一个元素是否为空
        String key = iterator.next();
        String song = songMap.get(key);        不再需要强制类型转换了
            System.out.println(key + "." + song);  //将 key 和 value 显示
        }
    }
    public void delete(String id){
        songMap.remove(id);
    }
}
```

执行效果：

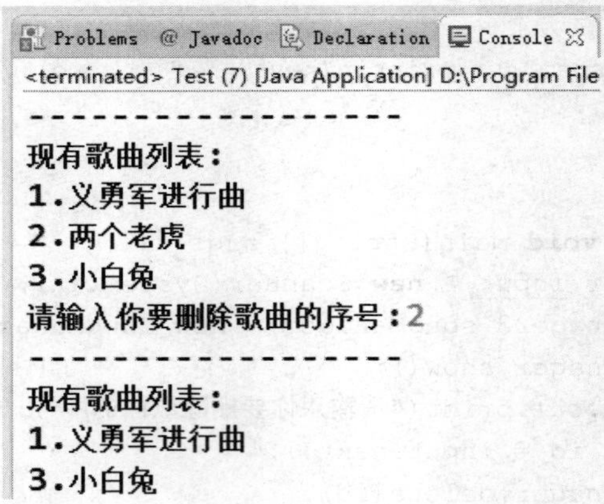

```
Problems  @ Javadoc  Declaration  Console
<terminated> Test (7) [Java Application] D:\Program File
-------------------
现有歌曲列表：
1.义勇军进行曲
2.两个老虎
3.小白兔
请输入你要删除歌曲的序号：2
-------------------
现有歌曲列表：
1.义勇军进行曲
3.小白兔
```

✧ 调试运行

利用泛型可以指定集合中的数据类型，此时集合中只能存储该数据类型。

示例代码：

```
Map<String,String> songMap = new HashMap<String,String>();
```

此时已经指定 key 和 value 均为 String 类型数据，否则将会报错，如图 4-5 所示。

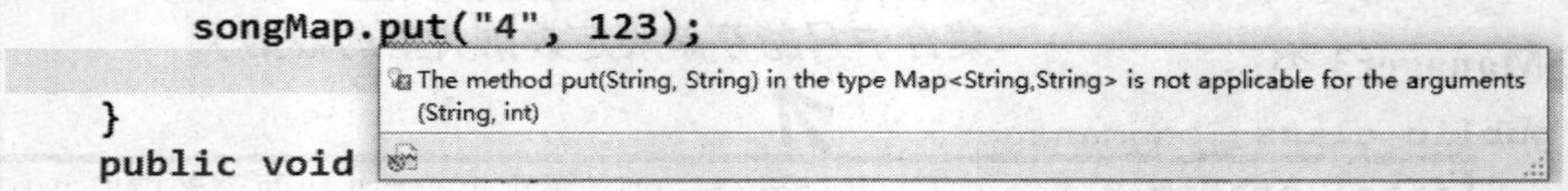

```
    songMap.put("4", 123);
}   The method put(String, String) in the type Map<String,String> is not applicable for the arguments
    (String, int)
public void
```

图 4-5 存入数据类型不符

当从集合中提取数据时，也就不再需要强制类型转换了。

✧ 维护升级

现在利用泛型实现显示歌曲详细信息功能。

示例代码：

```
SongManager4 类:
public class SongManager4 {
    List<SongDetail> songList = new ArrayList<SongDetail>();
                                                        //初始化集合
    /**
     * 初始化数据
     */
    public SongManager4() {
        songList.add(new SongDetail("义勇军进行曲",
"中国人民解放军军乐团", "中国军乐进行曲 20 首"));
    songList.add(new SongDetail("两个老虎", "群星", "如歌童年"));
    songList.add(new SongDetail("小兔子乖乖", "群星", "如歌童年"));
    }

    public void show() {
        System.out.println("------------------");
        System.out.println("现有歌曲列表:");
        Iterator<SongDetail> iterator = songList.iterator();
                                                // 得到 set 集合的迭代器
        int i = 1;  //用于显示歌曲前面的序号
        while (iterator.hasNext()) { // 判断下一个元素是否为空
            SongDetail songDetail = iterator.next();
                                                // 移动到下一个元素
                                                //并获取其中存放的数据
            System.out.println(i + "." + songDetail.getTitle());
                                                // 显示歌曲名称
            i++;
        }
    }

    public void showDetail(int id) {
        SongDetail songDetail = songList.get(id-1);
                                                //从集合中取出一首歌

        System.out.println("歌曲名称\t\t 歌手\t\t 所属专辑");
        System.out.println(songDetail.getTitle() + "\t\t"
+songDetail.getSinger()+ "\t\t" + songDetail.getAlbum());

    }
}
```

测试类：

```
public static void main(String[] args) {
        Scanner input = new Scanner(System.in);
        SongManager4 songManager = new SongManager4();
        songManager.show();
        System.out.print("请输入需要显示详细信息的歌曲编号:");
        int id = input.nextInt();
        songManager.showDetail(id);
}
```

执行效果：

```
Problems  @ Javadoc  Declaration  Console  Progress
<terminated> Test (7) [Java Application] D:\Program Files (x86)\Java\jre8\bin\ja
--------------------
现有歌曲列表:
1.义勇军进行曲
2.两个老虎
3.小兔子乖乖
请输入需要显示详细信息的歌曲编号:3
歌曲名称            歌手            所属专辑
小兔子乖乖          群星            如歌童年
```

通过上面的例子可以看到，利用泛型可以非常方便地使用集合，而不需要考虑类型转换问题。

项目实训与练习

一、操作题

有一个学生类，属性：姓名、年龄、成绩。方法：showInfo(),显示该学生信息。按照下列要求完成集合操作：

（1）创建一个 List，在 List 中增加三个学生，基本信息如下：

姓名	年龄	成绩
zhang3	18	78
li4	21	99
wang5	22	59

（2）在 li4 之前插入一个学生，信息为：姓名：zhao6，年龄：19，成绩：89。

（3）删除 wang5 的信息。

（4）利用迭代遍历，对 List 中所有的学生调用 showInfo()方法。

二、选择题

1．下列哪些方法是在 Collection 接口中定义的？（　　）

A．iterator()　　B．isEmpty()　　C．toArray()　　D．setText()

2. 如果希望数据有序存储并且便于修改，可以使用哪种 Collection 接口的实现类？（　　）

A．LinkedList　　B．ArrayList　　C．TreeMap　　D．HashSet

3．如果希望数据有序存储并且便于查询，可以是哪种Collection接口的实现类？（　　）

A．LinkedList　　B．ArrayList　　C．TreeMap　　D．HashSet

4．下列代码的运行结果是（　　）。

```
import java.util.*;
public class Test {
    public static void main(String[] args) {
        List<Integer> list = new ArrayList<Integer>();
        Iterator<Integer> it = list.iterator();
        System.out.println(it.next());
    }
}
```

A．0　　B．抛出异常　　C．编译错误　　D．运行错误

项目 5

员工信息管理程序

项目目标

了解 JDBC 的基本概念，熟练使用 JDBC API 提供的主要接口和类，掌握 JDBC 编程的基本步骤，利用 JDBC 技术实现对不同类型数据库（access，SQL Server，MySql）的操作，掌握 JDBC 编程在实际项目中的应用，为 Java 数据库系统开发打下良好的基础。

项目内容

以员工管理系统为例，详细讲解使用 JDBC 技术完成数据库表中记录的增、删、改、查操作。

任务 1　查询员工信息

✧ 需求分析

本程序利用 Java GUI 编程做出一个前台查询界面，按照员工表中的员工编号字段查询员工信息，后台利用 access 数据库管理系统建立数据库和员工表（见图 5-1）。程序实现的重点和难点在于如何使用 Java 语言进行数据库编程和配置 ODBC 数据源。

图 5-1　查询窗口

1．需求描述

在员工管理系统中，增加一个查询功能，按照员工编号字段查询员工的全部信息。程序后台需要利用 JDBC-ODBC 桥的方法连接 access 数据库，其中关键的地方在于如何配置数据源。打开 access2003 数据库管理系统，建立 Employee.mdb 数据库、 employee 表，表结构如表 5-1 所示。

表 5-1　员工信息 employee 表

字段名	数据类型	长度	含义	备注
eNum	文本	4	员工编号	主键
eName	文本	20	员工姓名	
eSex	文本	2	员工性别	
eAge	数字		员工年龄	
eDep	文本	20	员工部门	

在 employee 表中添加一些记录，如图 5-2 所示。添加记录完成后关闭 Access 数据库。

employee : 表

eNum	eName	eSex	eAge	eDep
1001	Ann	女	21	人事部
1002	高健琳	女	24	人事部
1003	刘海金	男	37	财务部
1004	赵楠楠	女	26	策划部
1005	林丽丽	女	28	财务部
1006	卢强	男	29	财务部
1007	李双	男	29	市场部
1008	杨雪	女	28	策划部

图 5-2　employee 表数据

2．运行结果

在 Eclipse 中运行该项目，结果如图 5-3 所示。

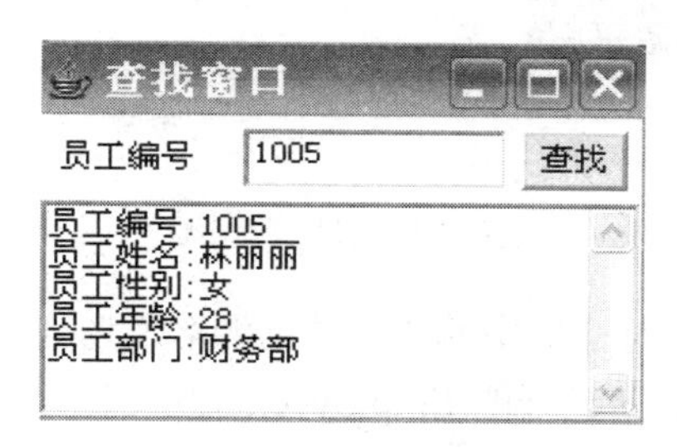

图 5-3　显示查询结果

✧ 知识准备

1．技能解析

利用 awt 编程实现了前台查询界面，使用 access 建立了数据库和对应的表，任务重点是如何使用 JDBC 编程将前台查询窗口和后台的数据库建立连接，以及如何建立、配置 ODBC 数据源，方法如下：

a．选择“开始菜单”|“控制面板”命令，双击“管理工具”|数据源（ODBC）图标。

b．在打开的“ODBC 数据源管理器”对话框中选择“系统 DNS”项，如图 5-4 所示。

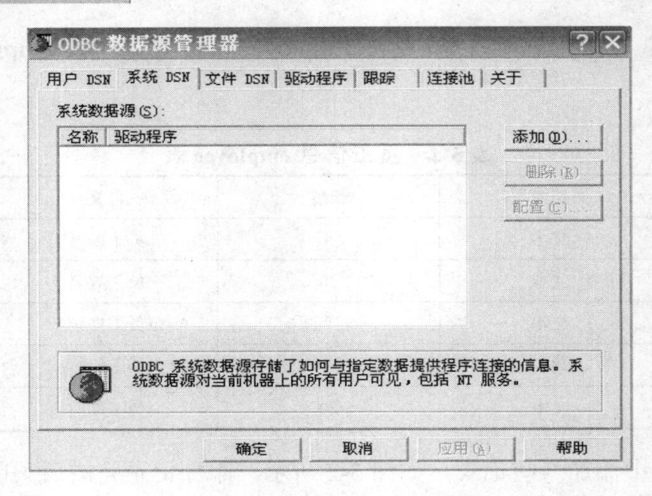

图 5-4　ODBC 数据源管理器

c．单击“添加”按钮，在弹出的对话框中选择“Microsoft Access Driver(*.mdb)”项，如图 5-5 所示。

图 5-5　设置数据源

d．单击“完成”按钮，在对话框中输入“数据源名”为 db（db 为用户自定义的数据名），然后单击“选择”按钮，选定创建好的 Employee.mdb 文件，如图 5-6 所示。

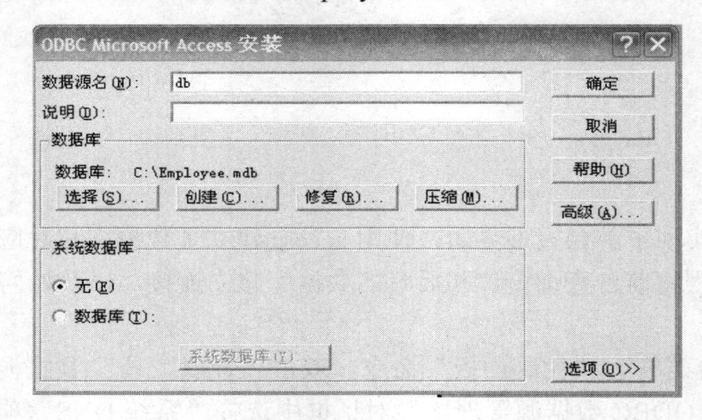

图 5-6　选定创建好的 Employee.mdb 文件

e．单击“确定”按钮，ODBC 数据源设置完成。

2．知识解析

前台主程序利用 JDBC_ODBC 类实现了查询窗口编程，该类需要继承 WindowAdapter 类，实现 ActionListener 接口，重写 actionPerformed 方法和 windowClosing 方法，把和数据库操作有关的代码写在 actionPerformed 方法中，因为该方法触发查询按钮，进行数据的查询操作。

JDBC API 提供的类和接口在 java.sql 包中定义（见表 5-2），这些类和接口提供了 Java 进行数据库编程的主要方法，熟练掌握这些类和方法是 JDBC 编程的主要学习目标。

表 5-2 JDBC API 包中定义的类和接口

类和接口名称	作用
java.sql.Driver	定义一个数据库驱动程序的接口
java.sql.DriverManager	用于 JDBC 驱动程序
java.sql.Connection	用于与特定数据库的连接
java.sql.Statement	Statement 的对象用于执行 SQL 语句并返回执行结果
java.sql.PreparedStatement	创建一个可以预编译的 SQL 对象
java.sql.ResultSet	用于创建表示 SQL 查询结果的结果集
java.sql.CallableStatement	用于执行 SQL 存储过程
java.sql.DatabaseMetaData	用于取得与数据库相关的信息如数据库、驱动程序、版本等
java.sql.SQLException	处理访问数据库产生的异常

✧ 编码实施

打开 Eclipes，创建一个名字为 Employee 的 Java Project，在该工程下新建名为 JDBC_ODBC 类。然后把下面的代码写入 JDBC_ODBC 类。

```
import java.awt.*;
import java.awt.event.*;
import java.sql.*;
public class JDBC_ODBC extends WindowAdapter implements ActionListener{
    Frame f=new Frame("查找窗口");
    Label l1=new Label("员工编号");
    TextField t1=new TextField(10);
    Button b1=new Button("查找");
    TextArea ta=new TextArea(10,10);
    Panel p1=new Panel();
    public void init(){
        f.add(p1,"North");
        f.add(ta,"Center");
        p1.add(l1);
        p1.add(t1);
        p1.add(b1);
        f.setVisible(true);
        f.pack();
        f.addWindowListener(this);
```

```
        b1.addActionListener(this);
    }
    public void windowClosing(WindowEvent e){
        System.exit(0);
    }
    public void actionPerformed(ActionEvent e){
        Connection conn=null;
        PreparedStatement pStmt=null;
        ResultSet rs=null;
        String url="jdbc:odbc:db";
        String DriverName="sun.jdbc.odbc.JdbcOdbcDriver";
        try{
            if(e.getSource()==b1){
                    String s1=t1.getText();
                    Class.forName(DriverName);
                    conn=DriverManager.getConnection(url);
            String sql="select * from employee where eNum='"+s1+"'";
                    pStmt=conn.prepareStatement(sql);
                    rs=pStmt.executeQuery();
                    while(rs.next()){
                        String eNum=rs.getString(1);
                        String eName=rs.getString(2);
                        String eSex=rs.getString(3);
                        int eAge=rs.getInt(4);
                        String eDep=rs.getString(5);
                        ta.append("员工编号:"+eNum+"\n");
                        ta.append("员工姓名:"+eName+"\n");
                        ta.append("员工性别:"+eSex+"\n");
                        ta.append("员工年龄:"+eAge+"\n");
                        ta.append("员工部门:"+eDep+"\n");
                            }
                    }

            }
        catch(Exception e1)
         {
            e1.printStackTrace();
         }
    }
    public static void main(String[] args) {
        new JDBC_ODBC().init();
    }
}
```

✧ 调试运行

程序运行时需要在员工编号文本框中输入正确的员工编号，单击查询按钮可以查询出该员工全部信息，单击关闭按钮退出查询窗口。

✧ 维护升级

该程序实现了根据某员工编号查询该员工全部信息的功能，也可以根据员工的其他字段进行查询，如根据员工姓名等。同学们也可以尝试为查询窗口增加查询员工全部信息功能，具体实现方法将在任务2中介绍。

任务2 查询全部员工信息

✧ 需求分析

在任务1的基础上，为员工管理系统增加查询全部员工信息功能，为了使同学们熟练掌握Java操作不同数据库的方法和步骤，任务2后台数据库使用Microsoft SQL Server 2008，介绍JDBC编程常用类和接口的使用。

1．需求描述

Microsoft SQL Server 2008作为后台数据库系统，建立一个名为Employees的数据库，数据库中包括employee表，表中有员工编号（eNum）、员工姓名（eName）、员工性别（eSex）、员工年龄（eAge）、员工部门（eDep）五个字段。表employee结构如图5-7所示。

表 - dbo.employee

列名	数据类型	允许空
eNum	varchar(8)	☐
eName	varchar(20)	☐
eSex	varchar(2)	☐
eAge	int	☐
eDep	varchar(30)	☐
		☐

图5-7 表employee结构

然后在该表中插入部分测试用数据，表employee中记录如图5-8所示。

eNum	eName	eSex	eAge	eDep
1001	Ann	女	21	人事部
1002	高健琳	女	24	人事部
1006	卢强	男	29	财务部
1003	刘海全	男	37	财务部
1004	赵楠楠	女	26	策划部
1005	林丽丽	女	28	财务部
1007	李双	男	30	市场部
1008	杨雪	女	28	策划部

图5-8 表employee记录

2．运行结果

在Eclipse中运行该项目，结果如图5-9所示。

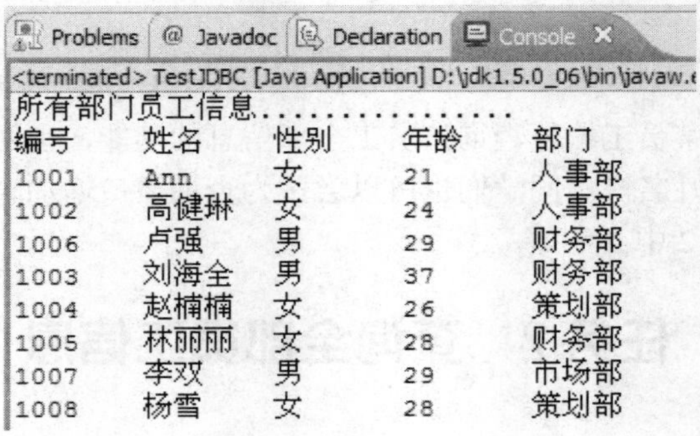

```
Problems  @ Javadoc  Declaration  Console ✕
<terminated> TestJDBC [Java Application] D:\jdk1.5.0_06\bin\javaw.e
所有部门员工信息.................
编号    姓名    性别    年龄    部门
1001    Ann     女      21      人事部
1002    高健琳  女      24      人事部
1006    卢强    男      29      财务部
1003    刘海全  男      37      财务部
1004    赵楠楠  女      26      策划部
1005    林丽丽  女      28      财务部
1007    李双    男      29      市场部
1008    杨雪    女      28      策划部
```

图 5-9　查询全体员工信息

✧ 知识准备

1．技能解析

在程序设计时，主要使用 DriverManager、Connection、Statement、PreparedStatement、ResultSet 等几个类和接口。它们之间的关系是：通过 DriverManager 类的相关方法能够建立同数据库的连接，建立连接后返回一个 Connection 类的对象，再通过该对象的方法创建 Statement 或 PreparedStatement 的对象，最后用 Statement 或 PreparedStatement 的方法执行 SQL 语句得到 ResultSet 类的对象，该对象包含了 SQL 语句的检索结果，通过这个检索结果可以得到数据库中的数据。

2．知识解析

连接数据库采用 DriverManager 类的 getConnection 方法，该方法返回 Connection 类对象。连接语句如下：

```
Connection  conn=DriverManager.getConnection(url, user,passwrod);
```

getConnection 方法有 3 个参数，url 表示要连接数据库的 url 地址，user 和 passwordwv 分别表示连接数据库的用户名和密码。

数据库的 url 一般格式为：

```
jdbc:drivertype:driversubtype://parameters
```

drivertype 表示驱动程序的类型，driversubtype 是可选的参数，parameters 通常用来设定数据库服务器的 IP 地址、端口号和数据库的名称。常见的几种数据库的 url 如下：

（1）用 JDBC-ODBC 桥连接的数据库，采用如下形式：

```
jdbc:odbc:datasource
```

其中 datasource 为 ODBC 数据源的名字。

（2）对于 SQL Server 采用如下形式：

```
jdbc:microsoft:sqlserver://localhost:1433;DatabaseName=MyDB
```

其中 MyDB 是用户建立的 SQL Server 数据库名字。

（3）对于 MySQL 数据库，采用如下形式：

```
jdbc:mysql://localhost:3306/MyDB
```

其中 MyDB 是用户建立的 MySQL 数据库名字。

（4）对于 Oracle 数据库，采用如下形式：

```
jdbc:oracle:thin@localhost:1521;orcl
```

其中 orcl 是数据库的 SID。

对数据库的操作，首先应该加载驱动程序类，JDBC 连接 SQL Server2005 的驱动类:com.microsoft.sqlserver.jdbc.SQLServerDriver，该类被封闭在 sqljdbc.jar 包中，所以应该在程序运行前将 sqljdbc.jar 包引入到项目中。方法是：在 Eclipse 的项目中单击右键，选择 properies 项，在弹出的对话框中选择左侧的 Java Build Path 项，然后在右侧的对话框中选择 Libraries 项,单击 Add External Jar 按钮，找到下载的 sqljdbc.jar 包所在的位置，最后单击 OK 按钮。

程序中的 Class.forName(DriverName)就是加载数据库驱动类，然后使用 DriverManager 类的 getConnection 方法进行数据库的连接。JDBC 连接不同数据库的步骤或是说方法几乎是相同的，不同的地方在于不同的数据库厂商所提供的数据库驱动类是不同的，还有就是连接数据库的 url 地址不同，以及登录数据库时所用到的用户名和密码是不同的。

✧ 编码实施

打开 Eclipes，创建一个名字为 Employee 的 Java Project，在该工程下新建名为 TestJDBC 类。然后把下面的代码写入 TestJDBC 类。

```
import java.sql.Statement;
import java.sql.Connection;
import java.sql.DriverManager;
import java.sql.ResultSet;
public class TestJDBC {
private static Connection conn=null;
private static Statement stmt=null;
private static ResultSet rs=null;
private final static String username="sa";
private final static String password="123456";
private final static String url="jdbc:sqlserver://localhost:1433;DataBaseName=
Employees";
    private final static String DriverName="com.microsoft.sqlserver.jdbc.
    SQLServerDriver";
    public static void main(String[] args) {
        try{
            Class.forName(DriverName);
            conn=DriverManager.getConnection(url,username,password);
            stmt=conn.createStatement();
            String sql="select * from employee";
            rs=stmt.executeQuery(sql);
            System.out.println("所有部门员工信息...............");
        System.out.println("编号"+"\t"+"姓名"+"\t"+"性别"+"\t"+"年龄
                        "+"\t"+"部门"+"\t");
            while(rs.next()){
            String eNum=rs.getString(1);
            String eName=rs.getString(2);
            String eSex=rs.getString("eSex");
            Integer eAge=rs.getInt("eAge");
```

```
                String eDep=rs.getString("eDep");
                System.out.print(eNum+"\t");
                System.out.print(eName+"\t");
                System.out.print(eSex+"\t");
                System.out.print(eAge+"\t");
                System.out.println(eDep);
                            }
                rs.close();
                stmt.close();
                conn.close();
                }
            catch(Exception e){
                e.printStackTrace();
            }
        }
    }
```

✧ 调试运行

运行该程序，发现不能正常显示出员工信息，并出现如图5-10所示的错误提示信息。

```
java.lang.ClassNotFoundException: com.microsoft.sqlserver.jdbc.SQLServerDriver
        at java.net.URLClassLoader$1.run(URLClassLoader.java:200)
        at java.security.AccessController.doPrivileged(Native Method)
        at java.net.URLClassLoader.findClass(URLClassLoader.java:188)
        at java.lang.ClassLoader.loadClass(ClassLoader.java:306)
        at sun.misc.Launcher$AppClassLoader.loadClass(Launcher.java:268)
        at java.lang.ClassLoader.loadClass(ClassLoader.java:251)
        at java.lang.ClassLoader.loadClassInternal(ClassLoader.java:319)
        at java.lang.Class.forName0(Native Method)
        at java.lang.Class.forName(Class.java:164)
        at TestJDBC.main(TestJDBC.java:18)
```

图 5-10 错误信息

原因是程序在运行中找不到一个名字叫 com.microsoft.sqlserver.jdbc.SQLServerDriver 的类，该类是使用 JDBC 连接 SQL2005 数据库的驱动类，所以需要将该类引入到项目中，具体方法是：下载一个名字叫 sqljdbc.jar 的包。com.microsoft.sqlserver.jdbc.SQLServerDriver 类就封装在此包中。接下来将该包加载到项目中，在 Eclipse 中右键项目名 Employee，选择 properies 项，在弹出的对话框中选择左侧的 Java Build Path 项，然后在右侧的对话框中选择 Libraries 项,单击 Add External Jar 按钮，找到下载的 sqljdbc.jar 包所在的位置，最后单击 OK 按钮。这样就将 jar 包引入到了项目中。如图 5-11 所示。

需要注意的是不同版本的数据库所需要的驱动类不一样，对应封装驱动类的jar包也不一样，如SQL Server 2005驱动类所对应的是sqljdbc.jar，SQL Server 2008驱动类所对应的是sqljdbc4.jar。

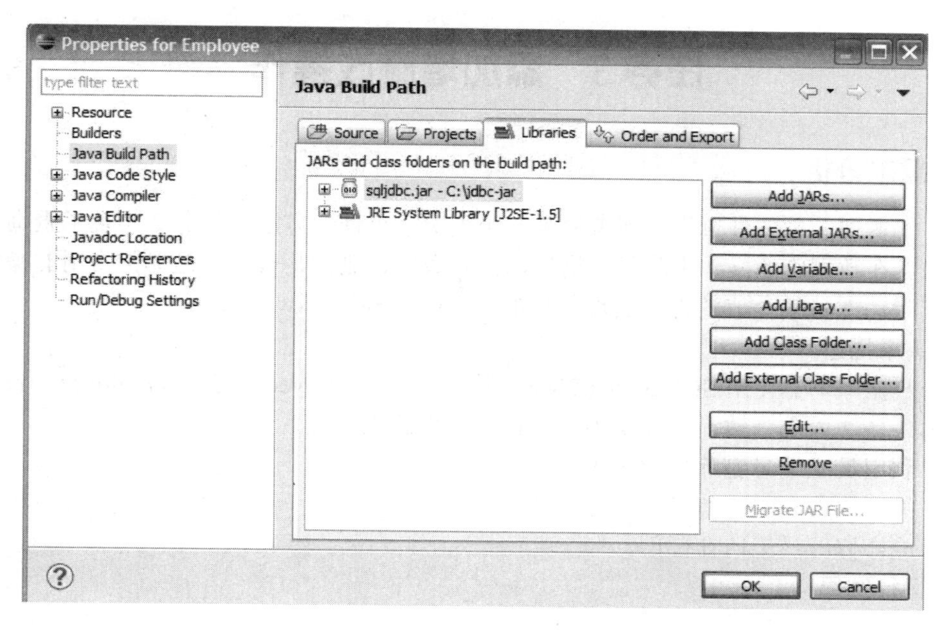

图 5-11　引入 jar 包

✧ 维护升级

对数据库进行一般查询使用 Statement 对象及其相关方法。数据库连接对象的 createStatement 方法能够返回 Statement 类的对象。具体方法为：

```
conn=DriverManager.getConnection(url,username,password);
stmt=conn.createStatement();
```

stmt 是 Statement 的类的对象，stmt 的 executeQuery(sql)方法用于执行 SQL 的查询语句，即 select 语句，执行结果将返回一个 ResultSet 类型的对象，该对象代表一个查询结果。

程序的查询结果可以通过 ResultSet 对象的相关方法获得。ResultSet 对象包含执行查询语句后得到的所有记录，记录的行号从 1 开始。一个 Statement 对象在同一时刻只能返回一个 ResultSet 对象。使用 ResultSet 的 next 方法，可以移动游标到下一记录，当游标指向记录末尾时，该方法将返回 false。通过 ResultSet 的 getXXX 方法（其中 XXX 表示 Java 的基本数据类型），可以取得某个字段的值。如：getString(int columnIndex)返回指定字段的字符串值，columnIndex 代表字段的序号；getInt(String columnName)返回指定字段的整型值，columnName 代表字段的名字。如果要访问 eNum 字段可以用以下语句：rs.getString("eNum"); 或 rs.getString(1);两种访问字段的方法结果是相同的。数据库操作结束后，要调用 Connection、Statement、ResultSet 类的 close 方法释放掉资源。

Scanner 类的 next 方法可以使用户从 System.in 中读取一个字符串，该类位于 java.util 包下，如果要读取的是整型数可以使用 Scanner 类的 nextInt 方法。

任务 3 添加增删改操作

✧ 需求分析

任务 1 实现了单条件查询，或者说是按照关键字查询的方法，任务 2 实现了查询全部信息的方法。在数据库的操作中除了查询外，插入数据、删除、修改都是非常常用的操作，任务 3 将利用动态参数传值的方法实现数据的增删改三种操作。

1．需求描述

后台数据库仍然使用 SQL Server 2008，数据库名 Employees，表名 employee，表中字段和数据与任务 2 内容一致。现在实现员工信息的增删改操作。

（1）利用员工编号删除指定员工。

（2）能够加入新员工信息。

（3）能够修改员工所在部门。

2．运行结果

运行结果如图 5-2 所示。

```
Problems  Javadoc  Declaration  Console  Progress
所有部门员工信息.................
编号    姓名    性别    年龄    部门
1001    Ann     女      21
1002    高建林  女      24
1003    刘海泉  男      37
1004    赵楠楠  女      26
1005    林丽丽  女      28
1006    卢强    男      29
1007    李双    男      30
1008    杨雪    女      28
*****************************************
---------------功能菜单--------------------
1、删除员工信息
2、添加员工信息
3、修改员工部门
请输入你要执行的功能:
```

图 5-12 查询结果

✧ 知识准备

1．技能解析

通过前两个任务的学习，利用 JDBC 实现数据库编程可以总结为以下几个步骤：

- 加载 JDBC 驱动程序。
- 获取连接接口。
- 创建 Statement 对象。
- 执行 Statement 对象。
- 查看返回的结果集。
- 关闭结果集对象。
- 关闭 Statement 对象。

- 关闭连接接口。

实现增删改操作时仍然按照这样的思路和步骤进行编程。在数据库的操作过程中经常需要多个条件来确定操作的数据，可以利用 PreparedStatement 类的 setString(int parameterIndex，String x）和 set（int parameterIndex，int x）方法动态将指定参数设置为给定 String 或 int 类型。实际应用中这种预编译提高了 SQL 执行效率，建议尽量使用 Prepared Statement 类代替 Statement 类。

2．知识解析

按照任务 3 的需求，利用 PreparedStatement 实现增删改的 SQL 语句表示成如下的形式：

删除：String sql = "delete from employee where eNum=?";语句中的？表示要动态获得的参数，利用给定的员工编号，删除员工。

插入：

修改：

String sql="select * from employee where eDep=? and eSex=?";语句中的“？”表示要动态获得的参数。PreparedStatement 接口继续了 Statement 接口，PreparedStatement 的对象可以执行带参数的查询语句，在执行查询语句前，用对象的 setXXX()方法设置参数的值，其中的 XXX 代表参数的数据类型。setString(1,eDepartment);方法中的第一个参数是 where 表达式中“？”代表的参数的序号，第二个参数是“？”代表的参数的值。

✧ 编码实施

打开Eclipes，创建一个名字为Employee的Java Project，在该工程下新建名为TestJDBC3类。然后把下面的代码写入TestJDBC3。

```
import java.sql.*;
import java.util.Scanner;

public class TestJDBC3 {
    private static Connection conn = null;
    private static PreparedStatement pStmt = null;
    private static ResultSet rs = null;
    private final static String uName = "sa";
    private final static String uPassword = "123";
    private final static String url =
            "jdbc:sqlserver://localhost:1433;DataBaseName=Employees";
    private final static String DriverName =
            "com.microsoft.sqlserver.jdbc.SQLServerDriver";

    public static void main(String[] args) {
        Scanner sc = new Scanner(System.in);
        try {
            Class.forName(DriverName);
            conn = DriverManager.getConnection(url, uName, uPassword);
            showAllEmployee();
    System.out.println("******************************************");
            System.out.println("--------------功能菜单--------------");
            System.out.println("1、删除员工信息");
            System.out.println("2、添加员工信息");
```

```
                System.out.println("3、修改员工部门");
                System.out.print("请输入你要执行的功能: ");
                int no = sc.nextInt();
                switch (no) {
                case 1:
                    delete();
                    break;
                case 2:
                    insert();
                    break;
                case 3:
                    update();
                    break;

                }
            } catch (ClassNotFoundException e) {
                // TODO Auto-generated catch block
                e.printStackTrace();
            } catch (SQLException e) {
                // TODO Auto-generated catch block
                e.printStackTrace();
            }finally{
                try {
                    rs.close();
                } catch (SQLException e) {
                    // TODO Auto-generated catch block
                    e.printStackTrace();
                }
                try {
                    pStmt.close();
                } catch (SQLException e) {
                    // TODO Auto-generated catch block
                    e.printStackTrace();
                }
                try {
                    conn.close();
                } catch (SQLException e) {
                    // TODO Auto-generated catch block
                    e.printStackTrace();
                }
            }

        }
    static void showAllEmployee() throws SQLException {
            String sql = "select * from employee";
            pStmt = conn.prepareStatement(sql);
            rs = pStmt.executeQuery();;
```

```
            System.out.println("所有部门员工信息................");
        System.out.println("编号" + "\t" + "姓名" + "\t" + "性别" + "\t" + "年龄"
                + "\t" + "部门" + "\t");
            while (rs.next()) {
                String eNum = rs.getString(1);
                String eName = rs.getString(2);
                String eSex = rs.getString("eSex");
                Integer eAge = rs.getInt("eAge");
                String eDep = rs.getString("eDep");
                System.out.print(eNum + "\t");
                System.out.print(eName + "\t");
                System.out.print(eSex + "\t");
                System.out.print(eAge + "\t");
                System.out.print(eDep + "\n");
            }

        }

    static void update() throws SQLException {
            Scanner sc = new Scanner(System.in);
            System.out.print("请输入员工编号:");
            int no = sc.nextInt();
            System.out.print("请输入该员工的当前部门: ");
            String depBefore = sc.next();
            System.out.print("请输入您要修改的部门名称: ");
            String newDep = sc.next();

            String sql = "update employee set eDep=? where eNum=? and eDep=?";
            pStmt = conn.prepareStatement(sql);
            pStmt.setString(1, newDep);
            pStmt.setInt(2, no);
            pStmt.setString(3, depBefore);
            pStmt.execute();
            showAllEmployee();

        }

    static void delete() throws SQLException {
            Scanner sc = new Scanner(System.in);
            System.out.print("请输入要删除的员工编号:");
            int no = sc.nextInt();

            String sql = "delete from employee where eNum=?";
            pStmt = conn.prepareStatement(sql);
            pStmt.setInt(1, no);
            pStmt.execute();
            showAllEmployee();
        }
```

✧ 调试运行

1．程序中 setString(1,x);方法中的第一个参数表示第几个"？"，第二个参数是"？"代表参数的值。同学们需要注意序号和参数的对应关系。

2．在上述程序中有如下代码：

```
pStmt.setInt(1, no);
```

该方法的第一个参数是从 1 开始的，如果第一个参数为 0，则会出现如下错误：

```
com.microsoft.sqlserver.jdbc.SQLServerException: 索引 0 超出范围。
    at com.microsoft.sqlserver.jdbc.SQLServerException.makeFromDriverEr
    at com.microsoft.sqlserver.jdbc.SQLServerPreparedStatement.setterGe
    at com.microsoft.sqlserver.jdbc.SQLServerPreparedStatement.setInt(U
    at chap05.jdbc3.TestJDBC3.delete(TestJDBC3.java:142)
    at chap05.jdbc3.TestJDBC3.main(TestJDBC3.java:30)
```

✧ 维护升级

请大家利用 PreparedStatement 类实现员工系统的添加员工信息功能，如图 5-13 所示。

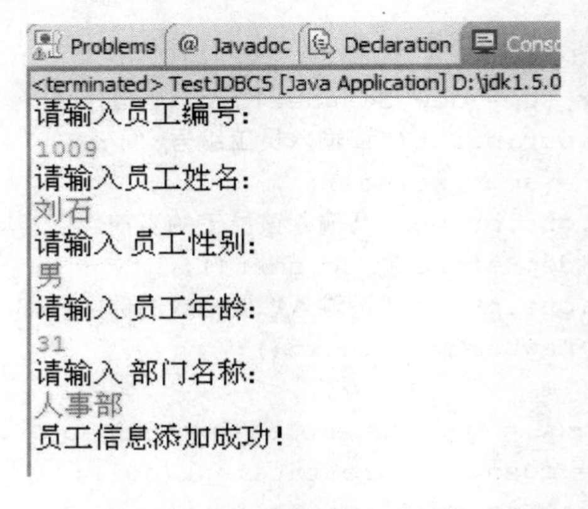

图 5-13 员工信息录入

参考代码如下：

```
String sql="insert into employee values(?,?,?,?,?)";
pStmt=conn.prepareStatement(sql);
pStmt.setString(1,eNum1);
pStmt.setString(2,eName1);
pStmt.setString(3,eMale1);
pStmt.setInt(4,eAge1);
pStmt.setString(5,eDepartment1);
int i=pStmt.executeUpdate();
System.out.println("员工信息添加成功！");
```

String sql="insert into employee values(?,?,?,?,?)";表示要向employee表中插入一条记录，并利用PreparedStatement接口的setXXX方法向要执行的SQL语句中5个"？"号动态的输入值。

项目实训与练习

一、选择题

1. Java 中，JDBC 是指（　　）。
   A. Java 程序与数据库连接的一种机制
   B. Java 程序与浏览器交互的一种机制
   C. Java 类库名称
   D. Java 类编译程序
2. 在利用 JDBC 连接数据库时，为建立实际的网络连接，不必传递的参数是（　　）。
   A. URL　　B. 数据库用户名　　C. 密码
3. JDBC 的模型对开放数据库连接(ODBC)进行了改进,它包含（　　）。
   A. 一套发出 SQL 语句的类和方法
   B. 更新表的类和方法
   C. 调用存储过程的类和方法
   D. 以上全部都是
4. JDBC 中要显式地关闭连接的命令是（　　）。
   A. Connection. close()　　B. RecordSet. close()
   C. Connection. stop()　　D. Connection. release()

二、填空题

1. JDBC APl 的含义是 Java 应用程序连接________的编程接口。
2. JDBC 驱动管理器使用________来装载合适的 JDBC 驱动。
3. Java 应用程序通过 JDBC. APl 向 JDBCDriverManager 发出请求，指定要装载的 JDBC 驱动程序代码，指定要连接的数据库的具体类型（品牌和版本号）和实例。JDBC.APl 主要是定义在________中的类和方法。
4. JDBC 的类都被汇集在________包中，在安装 JavaJDKl. 1 或更高版本时会自动安装。
5. 查询数据库的 7 个标准步骤是：载入 JDBC 驱动器、定义连接的网址 URL、建立连接、建立声明对象、执行查询或更新、处理结果、________。

三、实训题

1. 请编写程序，实现录入、查询和修改学生成绩的功能。
2. 请编写程序，设计并实现图书增加、修改、删除、查询的书籍管理系统。

项目 6

图形用户界面设计

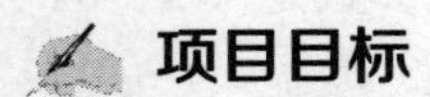

项目目标

了解 Swing 组件的基础知识，熟悉图形界面编程步骤，掌握常用容器、Swing 组件和布局管理器的创建与设置，理解事件处理机制，掌握对组件进行事件处理的方法；最终通过学习会创建各种组件并添加到容器中，能使用合适的布局管理器合理组织容器中的组件，会根据需求对组件添加事件处理。

项目内容

使用 swing 基本组件和布局管理器制作员工信息系统的注册界面，添加员工信息系统的事件处理功能，实现员工信息系统主界面，设计能够嵌入网页上的 Applet 程序。

任务1　用户注册界面设计

✧ 需求分析

一个良好设计的系统界面是整个系统的门面，美观、易操作的系统界面可以提高用户的工作效率，使用户乐于使用，方便操作。系统界面包括系统的登录注册界面、主界面、各种管理操作中的界面、操作过程的一些弹出对话框提示信息等。本次任务是通过 java 的图形用户界面设计工具容器和各种常用 Swing 组件，以及组件的布局设置器来实现员工信息系统的界面设计。

1．需求描述

通过用户注册界面完成员工信息系统的注册新用户功能。用户注册界面中有标签、按钮、文本框、复选框、密码框和作为性别选择的单选按钮。当用户填写好正确信息后，单击注册按钮，系统将把当前用户信息保存至数据库，为了保证程序的完整性，我们暂时显示一个简单的注册窗口信息，点击注册按钮后触发的事件响应代码在后续任务中讲解。

2．运行结果

用户注册界面设计如图 6-1 所示。

✧ 知识准备

本任务主要介绍图形界面编程的基础知识，用于组织各种组件的容器的创建和使用，以及在容器中添加基本组件后，组件的合理布局及事件响应机制。

图 6-1 用户注册界面设计

6.1.1 组件概述

图形用户界面（Graphical User Interface，GUI）是用于用户和程序之间进行交互的图形化用户界面，它是应用程序提供给用户操作的图形界面，包括窗口、按钮、菜单、工具栏和其他各种屏幕元素等，用户通过操作界面上的元素和鼠标共同完成对计算机发出指令、启动应用程序等操作任务。

AWT 是 Java1.0 提供的抽象视窗工具包，实现构建一个通用的 GUI，但受操作系统不同的影响，在不同平台上显示图形界面组件显示结果也有所不同。在 Java1.2 中及其后续版本中，又结合 AWT 的优点，用 100%纯 java 代码编写出完善、稳定的 Swing，于是 Swing 技术成为 Java GUI 编程的核心部分。Swing 在 Java1.2 版本中正式加入标准类库中，成为 Java 基础类库（JFC）的一部分。

Java 的 GUI 编程最重要的概念是容器和组件，组件是 GUI 的最基本组成部分，组件是指容器中的元素，以图形化的方式显示在屏幕上并能与用户进行交互的对象，如一个标签、一个按钮。GUI 编程时一般先添加容器，然后在容器中加入各种组件。

Swing 组件大都是 AWT 的 Container 类的直接子类和间接子类，各类之间的关系如下：

在 Java 图形界面编程时，一般要导入相应的包，AWT 的组件主要在 java.awt 包中，Swing 的组件主要在 java.swing 包中，事件处理类主要在 java.event 包中。

6.1.2 java.awt 包

java.awt 包是 Java 语言用来设计 GUI 的基本元素的类库，它包括了许多界面元素和资源，主要在三个方面提供界面设计支持：低级绘图操作、图形界面组件和布局管理以及界面用户交互控制和事件响应的类和接口。因此 AWT 相当复杂而且非常庞大，以致于被编制为一个主包（java.awt）和四个辅助包（java.awt.event、java.awt.image、java.awt.datatransfer 和 java.awt.peer）。

AWT 由下列包所组成，如图 6-2 所示。

java.awt 包中最核心的类就是 Component 类，它是构成 Java 图形用户界面的基础，大部分组件都是由该类派生出来的。Component 类是一个抽象类，其中定义了组件所具有的一般功能：基本的绘画支持（paint, repaint, update 等）、字体和颜色等外形控制（setFont, SetForeground 等）、大小和位置控制（SetSize, SetLocation 等）、图像处理（实现接口 ImageObserver）以及组件状态控制（SetEnable, isEnable, isVisible, isValid）等。

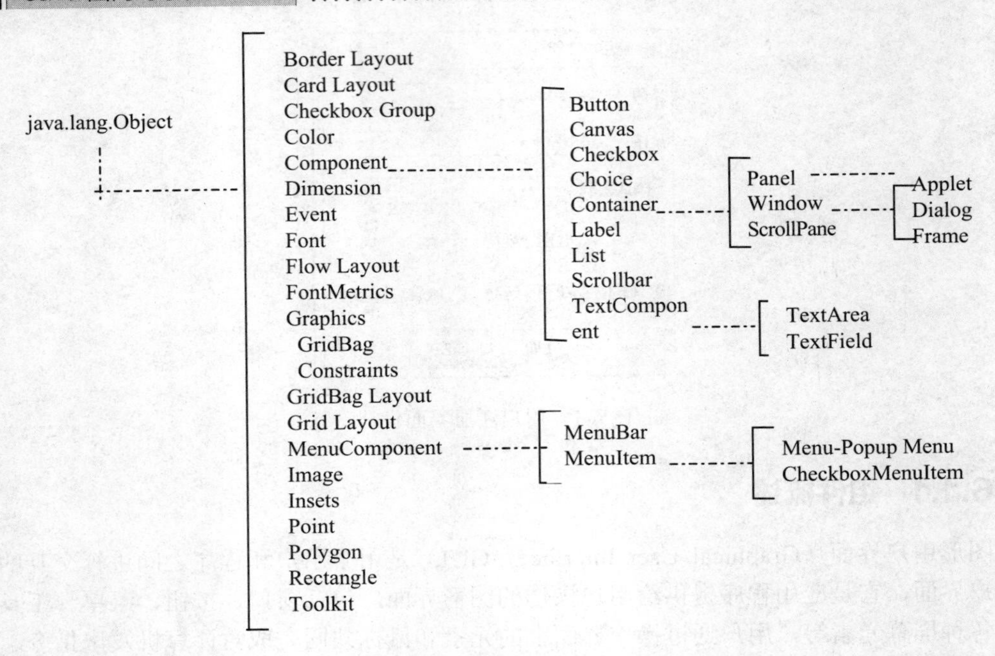

图 6-2 java.awt 包

AWT 中的组件分为容器（Container）类和非容器类两大类。容器本身也是组件，但其中可以包含其他的组件，也可以包含其他的容器。非容器类组件种类很多，比如按钮、标签等。Java 要显示的 GUI 组件都是 java.awt.Component 或 java.awt.MenuComponent 的子类。

常见的组件包括：Button, Checkbox, CheckboxGroup, Choice, Label, List, Canvas, TextComponent, Scrollbar，TextArea, TextField 等。

下面看一个基于 awt 包的传统窗体例子。

【例 6-1】创建一个 AWT 窗体。

```
Import java.awt.*;
public class awtWin
{
 public static void main(String args[])
 {
  Frame f=new Frame("This is a window!");
  f.setLocation(100,200);              //设置窗体显示位置
  f.setSize(200,400);                  //设置窗体尺寸
  f.setBackground(Color.red);          //设置窗体背景色
   FlowLayout fl=new FlowLayout();     //创建一布局管理器
  f.setLayout(fl);                     //设置窗体布局方式
  Button b1=new Button("Ok!");         //创建一个按钮组件
  f.add(b1);                           //向窗体中添加按钮，可直接用 add（ ）方法
  Button b2=new Button("Cancle!");
  f.add(b2);
  f.setVisible(true);
 }
}
```

说明：

（1）任何基于 AWT 包的传统窗体程序都必须导入相应的 AWT 包以方便编程。

（2）对于默认的新建窗体，左上角总是对齐屏幕的左上角，大小只能显示出标题栏，所以，必须设置常见的用于显示的属性。

（3）对于默认的新建窗体，并不直接显示，要想使窗体显示出来，必须设置 visible 属性为真。

（4）构造函数中的参数可以用于设置窗体的标题，也可以使用 setTitle()方法后期设置。

（5）添加组件的方法 add(Component)，如：add(Button1)。

6.1.3 java.swing 包

相对 AWT 来说，Swing 功能更强大、使用更方便，它的出现使得 Java 的图形用户界面上了一个新的台阶。Swing 中包含了丰富的类，它们都是 java.awt.Container 的子类,但包名已改为 javax.swing,它们中的大部分又是 javax.swing.JComponent 的子类。

除了那些常用组件如按钮、复选框和标签外，Swing 还包括许多新的组件，如选项板、滚动窗口、树、表格。同时，即使是常用组件，如按钮等，在 Swing 都增加了新功能。这些组件类名前都有一个 J，如 Jbutton、JLabel，对应的组件在 AWT 中称为 Button 和 Label。一般用轻量级（lightweight）这个术语描述这类组件。

Swing 组件大都是 AWT 的 Container 类的直接子类和间接子类，结构如图 6-3 所示。

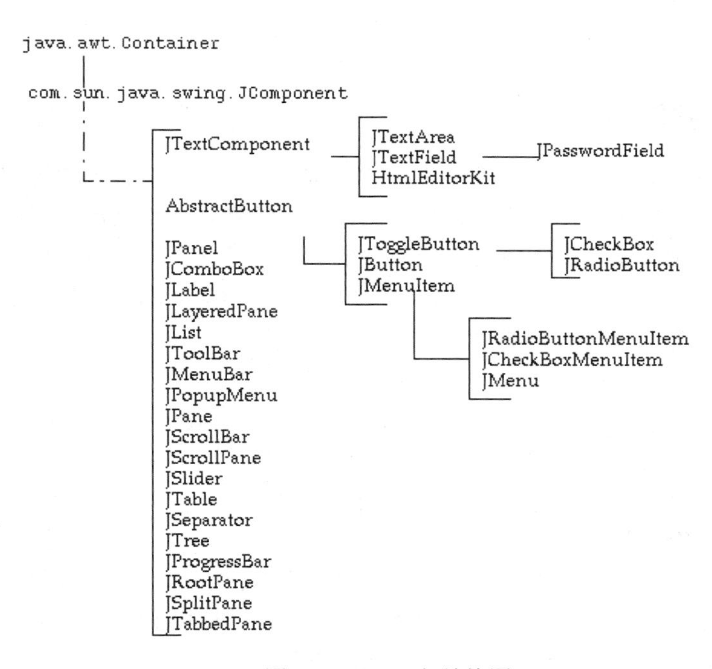

图 6-3 Swing 包结构图

JComponent 类继承自 java.awt.Container 类，是 Swing 组件的一个父类，封装了特定于 Swing 组件的基本特性和操作。

Swing 组件类型很多，按功能分类如下。

（1）顶层容器：有 Jframe（框架），JWindow（窗体），JDialog（对话框）和 JApplet（小

应用程序）。它们属于 swing 中的四个重量级组件。顶层容器的重要性在于，除了 Menubar 所有的 GUI 组件都需要被包含在顶层容器的一个默认容器 ContentPane 中。

（2）中间容器（又叫简单容器）：有 JPanel（面板）、JSplitPane（拆分面板）、JToolBar（工具栏）、JScrollPane（滚动面板）和 JTabbedPane（页签式面板）等。中间容器介于顶层容器与一般组件之间，是用来装别的组件用的，它是要被添加在顶层容器中的，能够使被它承载的组件能合适地、有组织地呈现出来。

（3）特殊容器：有 JLayeredPanel、 JRootPane InternalFrame 等，它们也是中间容器，但有着特殊的作用。

（4）基本组件：组件是一个可以以图形化的方式显示在屏幕上并能与用户进行交互的对象，例如 Jbutton、Jlabel、JList 等，是最基本图形用户界面的组成部分。

Swing 的超类 javax.swing.JComponent，继承自 java.awt.Component。Component 类中封装了组件通用的方法和属性，如图形的组件对象、大小、显示位置、前景色和背景色、边界、可见性等几个很有用的方法，它们被所有 JComponent 类的子类所继承，每个子类还有自己定义的方法，常用的方法有：

```
Add()                          //将组件添加到窗体中
    setEnable(boolean)         //设置是否可用
    setVisible(boolean)        //设置窗口是否可见
    setSize(int,int)           //设置宽和高
    getSize()                  //返回 Dimension 对象，其变 width 和 height 就是尺寸
    setText()                  //设置文本
    setValue()                 //设置数值
```

6.1.4 窗口容器类

容器（Container）是 Component 的子类，容器本身也是一个组件，不过它是用来容纳其他基本组件的组件。基本组件不能独立地显示出来，必须将基本组件放在特殊的组件（即容器）中才可以显示出来，而容器本身具有层次性的关系，就如同珠宝盒一样，大盒子里套小盒子，小盒子里还可以放更小的盒子，而珠宝就放在某一个盒子里，这里的珠宝就代表组件，盒子就代表容器。容器有助于在屏幕上安排布置 GUI 组件。

1．窗口类容器

Swing 提供了四个在屏幕上显示窗口的组件：JWindow、JFrame、JApplet 和 JDialog。这四个组件统称为最上层组件，它们相互之间的区别不明显。因为其余的 Swing 组件都必须依附在此四组件之一上才能显示出来，又称窗口类容器。

2．JFrame 和 JPanel

窗体类 JFrame 还叫框架类，是一种带有标题栏、菜单和边界的窗口，而且允许调整大小，是最重要的顶层容器之一，而且是应用程序构建框架时必须使用的顶层容器。JFrame 类是 AWT 中 Frame 类的子类，它还加入了一些 Swing 所独有的特性。

（1）构造方法：

```
JFrame()                       //构造一个新的不可见的窗体对象
JFrame(String   title)         //构造一个新的、不可见的、具有指定标题的窗体对象
```

（2）常用方法：

```
public void setBound(int x,int y,int w,int h)  //设置窗口左上角位置和窗口大小
```

```
public void setTitle(String title)              //设置窗口标题
public void setResizable(Boolen boolean)        //设置窗口是否可调大小
public void setLocation(int x,int y)            //设置窗口左上角位置
public void pack()                              //用紧凑方式自动设定窗口大小
JFrame 可以调用
```

public void setDefaultCloseOperation(int operation)设置单击窗体右上角的关闭图标后，程序应做出怎样的处理。其中参数 operation 取有效常量

JFrame. DO_NOTHING_ON_CLOSE：什么也不做。

JFrame. HIDE_ON_CLOSE：隐藏窗体，这是默认选项。

JFrame. EXIT_ON_CLOSE：结束程序。

在 GUI 编程时，经常通过继承 JFrame 类定义新的窗体类，在新的窗体类中创建各种组件并添加到新窗体中。

JPanel（面板）是一种常用的容器，它常常用做中间容器，即在面板中添加组件，然后再将面板添加到其他容器。这样一方面可以将图形用户界面的组件进行分组，另一方面还可以形成较合理的组件布局方式。面板还常常用做 swing 图形用户界面的画板。

下面看一个基于 Swing 的窗体例子。

【例 6-2】JFrame 与 JPanle 简单示例。

```
import javax.swing.*;
public class swingWin
{
 public static void main(String args[])
 {
  JFrame f=new JFrame("Swing 窗口示例");
  JLabel hy=new JLabel("欢迎使用员工信息系统！");
  JPanel jp=new JPanel();
     f.setLocation(100,200);
  f.setSize(300,200);
  jp.add(hy);
  f.add(jp);
 f.setDefaultCloseOperation(JFrame.EXIT_ON_CLOSE);
  f.setVisible(true);
 }
}
```

运行效果结果如图 6-4 所示。

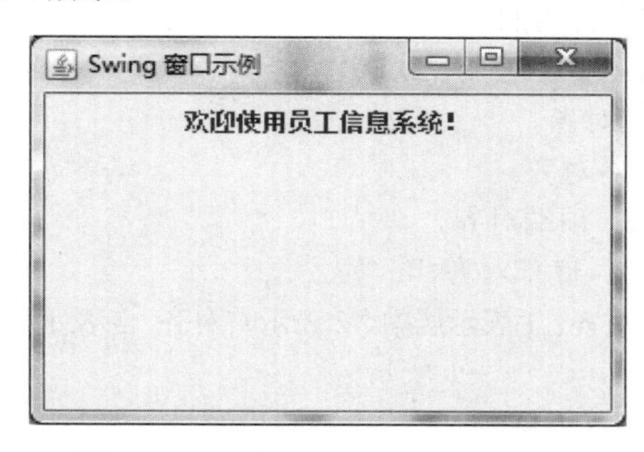

图 6-4 JFrame 与 JPanle 简单示例结果

这个程序先是生成一个窗体的实例，然后设置位置和大小，并且向窗体中添加了一个JLabel 标签组件。它是一个具有标题、大小及背景颜色的窗体。

6.1.5 容器的布局

在设计一个 GUI 界面时，我们会在各窗口容器上添加若干组件，如果只是简单添加，一旦改变窗口大小时，组件的排列会发生混乱。那么如何设置各种组件在容器中的摆放方式，使显示界面能够自动适应不同分辨率的屏幕而实现理想布局？

为了使容器内各个组件的位置摆放合理，Java 采用的是用“布局管理器”进行相对定位的方法。所有容器都有一个默认的布局管理器，它负责管理组件的排列顺序、组件的大小、位置，以及当窗口移动或调整大小后组件如何变化等。

Java 常用的几种布局管理器有：

FlowLayout——流式布局

BorderLayout——边界布局

GridLayout——网格布局(GridLayout)

CardLayout——卡片布局

GridBagLayout——网格包布局

这些布局管理器仍然沿用 AWT 中的，因此在使用时导入包“java.awt.*”。

1．FlowLayout 流式布局

它的排版方式就像流程或文本处理器在处理一段文字一样，将所有组件从左到右依次排列，一行不够时将自动转到下一行继续排列。当改变窗口大小时，组件将重新排列，但组件大小不变。流式布局下组件的对齐方式有左对齐、右对齐和居中对齐等。它是 Panel、Applet 的缺省布局管理器，一般用来安排面板中的按钮或复选框的排列。

FlowLayout 类的构造函数：

```
FlowLayout( )                       //使用居中方式构造流式布局，并且组件间默认的水平和垂直间
                                      隔是 5
FlowLayout(int align)               //使用给定的对齐方式构造流式布局，并且组件间默认的水平和
                                      垂直间隔是 5
FlowLayout(int align, int hgap, int vgap)      //使用给定的对齐方式构造流
                                                 式布局，并且指定组件间垂直
                                                 和水平间隔
```

构造函数的三个参数意义分别如下。

align：代表对齐方式，常用的有：

FlowLayout.CENTER：中心对齐（默认）。

FlowLayout.LEFT：左对齐。

FlowLayout.RIGHT：右对齐。

Flowlayout.LEADING：顶端对齐。

Flowlayout.TRAILING：底部对齐。

（或者用整数 0 表示左对齐，1 表示居中，2 表示右对齐，3 表示顶端对齐，4 表示底部对齐）

hgap 代表水平间距；

vgap 代表垂直间距。

FlowLayout 类的常用函数：

```
int  getAlignment( )                    //获得对齐方式
void setAlignment(int align)            //设置对齐方式
int getHgap()                           //获得组件间的水平间隔
void setHgap(int vgap)                  //设置组件间的水平间隔
int getVgap()                           //获得组件间的垂直间隔
void setVgap(int vgap)                  //设置组件间的垂直间隔
void removeLayoutComponent comp)        //从布局中移除指定的组件
```

创建了某个布局方式类的对象后，如何指定它的布局方式呢？

方法是调用容器类的 setLayout()方法来指定所需的布局方式。如：setLayout(new FlowLayout());

Java 中不希望采用绝对定位布局而采用的是相对定位布局，如需要使用绝对定位布局，则应该取消容器的当前布局方式 setLayout(null))，然后使用 setLocation() setSize() setBounds()设置组件的大小及位置。

【例 6-3】FlowLayout 流布局管理器举例。

```
import java.awt.FlowLayout;
import javax.swing.*;
public class FlowLayoutTest
{
 public static void main(String [] args)
 {
  JFrame f=new JFrame("流布局管理器示例");
  f.setLocation(10,100);
  f.setSize(240,200);
  FlowLayout fl =new FlowLayout();  //A 行  创建流式布局管理器
  fl.setVgap(5);
  fl.setHgap(15);
  f.setLayout(fl);
  JButton jb1=new JButton("我先来的");
  JButton jb2=new JButton("我第二");
  JButton jb3=new JButton("我第三");
  f.add(jb1);
  f.add(jb2);
  f.add(jb3);
   f.setVisible(true);
 }
}
```

运行效结果如图 6-5 所示。

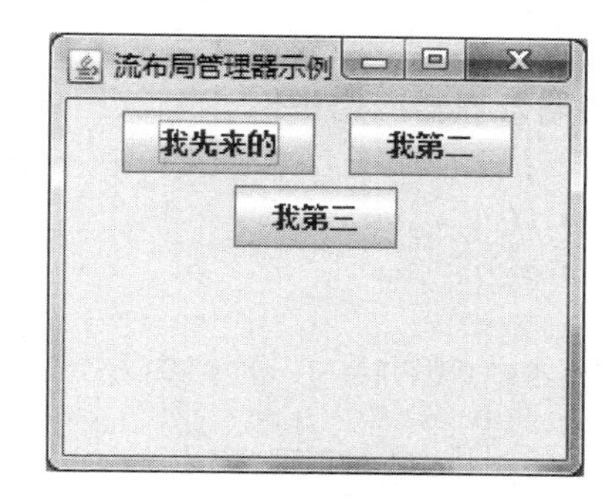

图 6-5　流式布局效果图

【程序分析】

（1）改变窗口的大小，那么组件的排列方式也会跟着变化。

（2）组件的顺序和组件的尺寸是确定的，不随窗口而改变。

当然，流式布局管理器类也可以定义很多布局属性，从而进行必要的一些样式调整，如：将上例的A行改成new FlowLayout(1,10,100);可以尝试着改变窗体的大小，观察按钮的布局，特别注意当一行显示不下而换行显示按钮时。中间的垂直间距和第一行按钮距窗体标题栏的间距都是100像素单位。但是，如果将窗体收缩太小，则可能无法显示全部组件。

2．BorderLayout 边界布局

Window、Frame、Dialog的默认布局为边界布局。这种布局将容器分成5个区域：North，South， West， East和Center，分别对应于窗口的顶部、底部、左部、右部和中部。每个区域只能放置一个组件，可以使用add（）方法将相关组件添加到指定的5个区域。

在使用BorderLayout布局的时候，如果容器的大小发生变化，其变化规律为：组件的相对位置不变，大小发生变化。例如容器变高了，则North、South区域不变，West、Center、East区域变高；如果容器变宽了，则West、East区域不变，North、Center、South区域变宽。

BorderLayout类的构造函数：

```
public Borderlayout()                           //构造一个边界布局管理器
public Borderlayout(int hgap,int vgap)          //构造一个边界布局管理器，并且指定组件之间的垂直和水平间隔
```

构造函数的参数 hgap代表水平间距；vgap代表垂直间距。

BorderLayout类的常用函数：

```
int getHgap()                              //获得组件间的水平间隔
void setHgap(int vgap)                     //设置组件间的水平间隔
int getVgap()                              //获得组件间的垂直间隔
void setVgap(int vgap)                     //设置组件间的垂直间隔
void removeLayoutComponent comp)           //从布局中移除指定的组件
```

在使用BorderLayout布局管理器时，需要注意以下几个问题：

组件添加的时候必须指定区域。添加方法为：

```
add(Object,String);                        //Object是被添加组件
```

String 为组件添加位置，可选 BorderLayout.NORTH，BorderLayout.SOUTH，BorderLayout.EAST，BorderLayout.WEST，BorderLayout.CENTER五者之一。

边界布局的局限性：单个区域不能显示多个组件。因为在边界布局中单个区域只能显示最后被添加的组件。解决这个问题的办法是：把组件添加至一个中间容器，再把中间容器添加到区域中。

【例6-4】BorderLayout布局管理器应用举例。

```
import javax.swing.*;
import java.awt.*;
public class BorderLayoutTest
{
 public static void main(String [] args)
 {
  JFrame f=new JFrame("边界布局管理器示例");
  f.setLocation(20,200);
  BorderLayout bl =new BorderLayout(5,5);
```

```
        f.setLayout(bl);
        JButton jb1=new JButton("上");
        JButton jb2=new JButton("下");
        JButton jb3=new JButton("左");
        JButton jb4=new JButton("右");
        JButton jb5=new JButton("中");
        f.add(jb1,BorderLayout.NORTH);
        f.add(jb2,BorderLayout.SOUTH);
        f.add(jb3,BorderLayout.EAST);
        f.add(jb4,BorderLayout.WEST);
        f.add(jb5,BorderLayout.CENTER);
        f.setVisible(true);
    }
}
```

运行效果如图 6-6 所示。

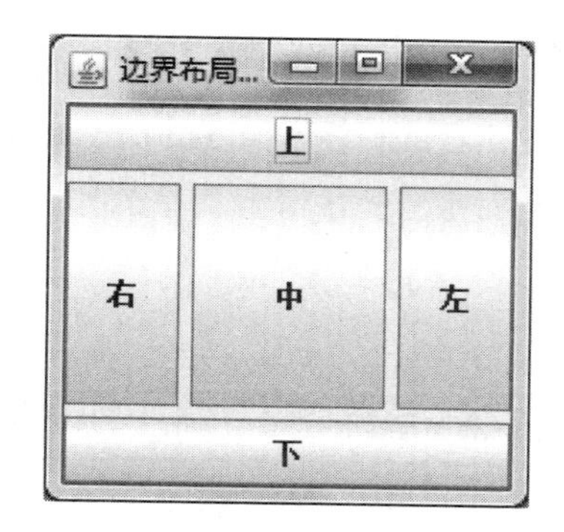

图 6-6 BorderLayout 边界布局示例

【程序分析】程序中定义了五个按钮，分别放置在 BorderLayout 布局的 5 个区域内，各个区域的间距为 5，每个按钮大小自动调整到各个区域的大小，此窗口有以下特点：

（1）改变窗口的大小，组件的排列方式保持不变。

（2）组件的大小会随窗口而改变。

（3）NORTH 与 SOUTH 有确定的高度，WEST 与 EAST 有确定的宽度，实际编程时，不一定所有的区域都有组件，如果四周的区域（West、East、North、South 区域）没有组件，则由 Center 区域去补充，但是如果 Center 区域没有组件，则保持空白，应试着修改程序观察效果。

3．GridLayout 网格布局

GridLayout 的布局管理格式是将版面分割成行数（rows）×列数（columns）的网格状版面。这样各个组件就可以按行列放置到每个网格中，每个组件的大小一样。在向 GridLayout 添加组件时，其顺序是从网格的左上角开始，从左向右排列，直到排满一行，再从下一行开始从左向右依次排列。GridLayout 类的构造函数如下：

GridLayout 类的构造函数：

```
public GridLayout ()                          //构造一行一列的网格布局
public GridLayout (int row,int columns)       //构造具有指定行数和列数的网格布局
public GridLayout (int row,int columns,int hgap,int vgap)
                    //构造具有指定行数和列数的网格布局,并指定组件间的水平垂直间隔
```

```
GridLayout类的常用函数:
init getRows()                              //获得此布局的行数
void  setRows(int rows)                     //设置此布局的行数
init getColumns  ()                         //获得此布局的列数
void  setColumns  (int cols)                //设置此布局的列数
int getHgap()                               //获得组件间的水平间隔
void setHgap(int vgap)                      //设置组件间的水平间隔
int getVgap()                               //获得组件间的垂直间隔
void setVgap(int vgap)                      //设置组件间的垂直间隔
void removeLayoutComponent comp)            //从布局中移除指定的组件
```

【例6-5】GridLayout布局管理器应用举例。

```
import java.applet.Applet;
import java.awt.*;
public class exa6_5  extends Applet{
    public void init(){
        setLayout(new GridLayout(3,3));
        for(int i=1;i<=8;i++){
        add(new Button("Button" + i));  //Applet容器内可直接添加组件
        }
    }
}
```

运行效果如图6-7所示。

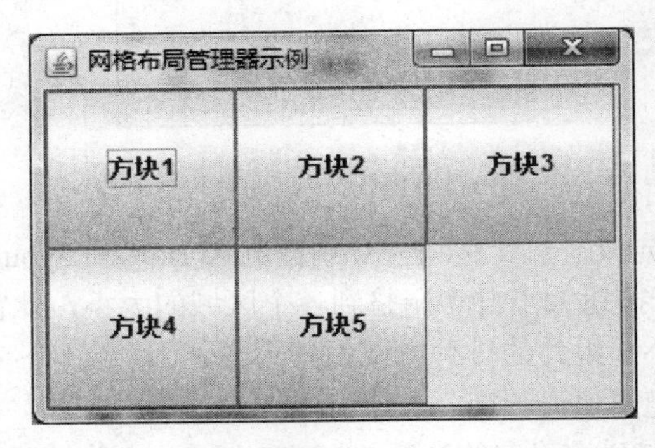

图6-7　网格布局效果图

【程序分析】这种布局下窗口的特点：

（1）改变窗口的大小，组件的排列方式保持不变，组件之间的行列间距保持不变。

（2）所有组件的宽度和高度相同，组件的大小会随窗口的尺寸而相应改变。

其他布局管理器还有如卡片布局（CardLayout）、网格袋布局（GridBagLayout）等，功能非常强大，使用时也比较复杂，考虑到一般的读者很少会使用到这种布局管理器，这里不做更多的介绍，读者可以在JDK文档中看到其详细说明及案例子程序。

✧ 编码实施

注册界面实现代码如下：

```
import java.awt.Dimension;
import java.awt.FlowLayout;
```

```
import java.awt.Toolkit;
import java.awt.event.ActionEvent;
import java.awt.event.ActionListener;

import javax.swing.ButtonGroup;
import javax.swing.JButton;
import javax.swing.JCheckBox;
import javax.swing.JFrame;
import javax.swing.JLabel;
import javax.swing.JPanel;
import javax.swing.JPasswordField;
import javax.swing.JRadioButton;
import javax.swing.JTextField;

public class RegPanel extends JFrame {
    public static final int WIDTH = 250;
    public static final int HEIGHT = 260;
    public RegPanel(){
        setTitle("新用户注册");
        JPanel panel = new JPanel(new FlowLayout(FlowLayout.CENTER));

        JPanel p1 = new JPanel();
        JLabel label1 = new JLabel("用  户  名");
        final JTextField userName = new JTextField(15);
        p1.add(label1);
        p1.add(userName);
        panel.add(p1);

        JPanel p2 = new JPanel();
        JLabel label2 = new JLabel("密码");
        final JPasswordField password = new JPasswordField(15);
        p2.add(label2);
        p2.add(password);
        panel.add(p2);

        JPanel p3 = new JPanel();
        JLabel label3 = new JLabel("确认密码");
        final JPasswordField rePassword = new JPasswordField(15);
        p3.add(label3);
        p3.add(rePassword);
        panel.add(p3);

        JPanel p4 = new JPanel();
        JLabel sex = new JLabel("性别");
        final ButtonGroup group = new ButtonGroup();
        final JRadioButton male = new JRadioButton("男");
        male.setSelected(true);
```

```
        final JRadioButton female = new JRadioButton("女");
        group.add(male);
        group.add(female);
        p4.add(sex);
        p4.add(male);
        p4.add(female);
        panel.add(p4);

        JPanel p5 = new JPanel();
        JLabel label5 = new JLabel("感兴趣的领域");
        final JCheckBox love1 = new JCheckBox("运动");
        final JCheckBox love2 = new JCheckBox("旅游");
        final JCheckBox love3 = new JCheckBox("美食");
        p5.add(label5);
        p5.add(love1);
        p5.add(love2);
        p5.add(love3);
        panel.add(p5);

        JPanel p6 = new JPanel(new FlowLayout(FlowLayout.CENTER));
        JButton reg = new JButton("注册");
        JButton clear = new JButton("清空");
        p6.add(reg);
        p6.add(clear);
        panel.add(p6);
        this.add(panel);

        Dimension d = Toolkit.getDefaultToolkit().getScreenSize();
        int x = (int) ((d.getWidth()-WIDTH)/2);
        int y = (int) ((d.getHeight()-HEIGHT)/2);
        setBounds(x, y, WIDTH, HEIGHT);
        setDefaultCloseOperation(EXIT_ON_CLOSE);
        setVisible(true);
        setResizable(false);
    }
    public static void main(String[] args) {
        new RegPanel();
    }
}
```

✧ 调试运行

1. 创建基本窗体的步骤：

- ✓ 导入 swt 和 swing 包；
- ✓ 继承 JFrame 类；
- ✓ 在类中定义组件；
- ✓ 在构造方法中创建组件；

✓　在构造方法中添加组件；
✓　设置窗体属性；
✓　显示窗体；
✓　在主函数中创建对象；
✓　所有布局管理器都可以添加任意组件。

2．在 Swing 引入 setDefaultCloseOperation()之前，AWT 关闭窗口必须使用程序监视窗口，获得相关操作并处理和做出响应（使用事件处理类 java.awt.event 包）。

3．设置组件的位置和大小的方法：

setBounds (int x, int y, int width, int height) 方法相当于：setLocation(int x,inty) 和 setSize (int width, int height)。

4．将组件的位置 Location 设置为：水平居中，垂直居中。

```
Dimension screenSize=Toolkit.getDefaultToolkit().getScreenSize();
setLocation((screenSize.width-getWidth())/2,(screenSize.height-getHei
ght())/2);
```

✧ 维护升级

Swing 还提供了一些高级布局管理器，如箱式布局管理器、卡片布局管理器、网格组布局管理器以及弹簧布局管理器，通过使用这些布局管理器，可以设计出更好、更适用的程序界面。

箱式布局管理器 BoxLayout 用来管理一组水平或垂直排列的组件。如果是用来管理一组水平排列的组件，则称为水平箱；如果是用来管理一组垂直排列的组件，则称为垂直箱。

BoxLayout 类仅提供了一个构造方法 BoxLayout（Container target,int axis），其入口参数 target 为要采用该布局方式的容器对象；入口参数 axis 为要采用的布局方式，如果将其设置为静态常量 X_AXIS，表示创建一个水平箱，组件将从左到右排列，设置为静态常量 Y_AXIS 则表示创建一个垂直箱，组件将从上到下排列。

无论水平箱还是垂直箱，当将窗体调小至不能显示所有组件时，组件仍会排列在一行或一列，组件按照添加到容器中的先后顺序进行排列。默认情况下，由箱式布局管理器实现的组件之间没有间距，如果要在组件之间设置间距，可以通过使用 Box 类提供的 6 个不可见组件实现，这些组件就是专门用来设置箱式布局管理器的。

任务 2　添加员工信息系统的事件处理

✧ 需求分析

用户注册界面中有标签、按钮、文本框、复选框、密码框和作为性别选择的单选按钮。当用户填写好正确信息后，单击注册按钮，系统将把当前用户信息保存至数据库，为了保证程序的完整性，我们暂时显示一个简单的注册窗口信息，点击注册按钮后触发的事件响应代码在后续任务中陆续讲解。

1．需求描述

通过用户注册界面完成员工信息系统的注册新用户功能。

2．运行结果

运行结果如图 6-8 所示。

```
Problems  @ Javadoc  Declaration  Project Migration
RegPanelEve [Java Application] C:\Users\Administrator\AppData\l
新用户加入到数据库中
用户名：菜鸟一枚
密码：woyuanyi
重复密码：woyuanyi
性别：男
兴趣：运动 旅游
```

图 6-8 实现注册的事件处理

✧ 知识准备

图形用户界面程序是事件驱动的，比如我们希望当按下按钮后会引发相应的动作来响应它，这样组件和用户才能交互进行下去，要实现这种交互就要用到 java 的事件处理。对于 GUI 程序与用户操作的交互功能，Java 使用了一种自己专门的方式，称之为事件处理机制，java 采用了委托事件处理模式，即组件本身没有用成员方法来处理事件，而是将事件委托给事件监听者处理。

java.awt.event 包提供处理由 AWT 组件所激发的各类事件的接口和类。

6.2.1 事件处理模式

Java 采用“授权事件模型”（Event Delegation Model）的事件处理模式。其原理为：事先定义多种事件类型（即用户在 GUI 组件上进行的操作，如单击事件），当用户对某个 GUI 组件进行操作时，可能触发相应的事件，这个组件就是事件源。如果此组件注册了事件监听器（可以通过 addXXXXListener 方法注册），事件被传送给已注册的监听器，事件监听器负责处理事件的过程。一个组件可以注册一个或多个监听器。

这个模式涉及的主要概念如下。

（1）事件（Event）：它描述在用户界面的用户交互行为所产生的一种效果，以类的形式出现，例如：鼠标操作对应的事件类是 MouseEvent.

（2）事件源（Event Source）：产生事件的场所，是一种组件对象。

（3）事件处理方法（Event handler）：接收、解析和处理事件类对象，实现和用户交互的方法。

（4）事件监听器（Event　Listener）：调用事件处理方法的对象。

6.2.2 事件处理的实现原理

使用授权处理模式进行事件处理主要有如图 6-9 所示几步。

① 若要保证一个组件产生的某种类型事件 XXXEvent 有效，需要注册一个“事件监听器”监听它。用如下语句设置事件监听器：

事件源.addXXXListener（XXXListener 代表某种事件监听器）。

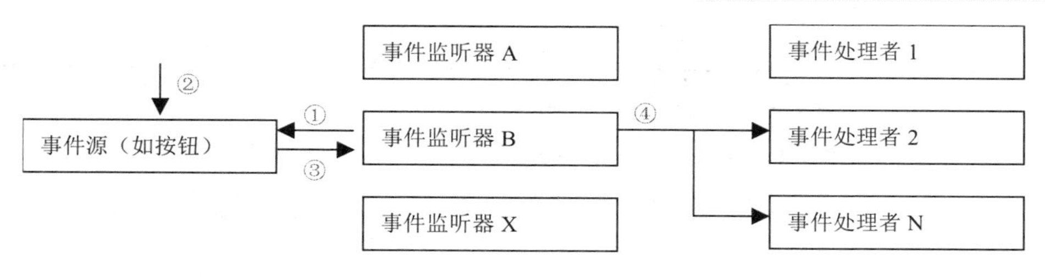

图 6-9 事件处理原理

② 组件作为事件源,不同类型的组件会产生特定类型的事件。

③ 一旦产生事件后，事件会被传送给已注册的一个或多个监听器。事件监听器是实现了与该事件相对应的 XXXListener 接口的类。

④ 事件监听器监听到事件后，会根据事件的类型调用相应的事件处理方法。

这样就可以处理图形用户界面中的对应事件了。

例如：当用户单击按钮后将触发动作事件 ActionEvent, 假设在这之前，我们已经使用 addActionListener(frame)方法为按钮注册了事件监听器 frame，那么按钮产生单击的动作事件后，将由事件接口 ActionListener 中的 actionPerformed（ActionEvent e）方法处理该事件。事件处理方法中常常要获知产生事件的事件源，用 getSource（）方法返回产生事件的事件源。要删除事件监听者可以使用语句：事件源.removeXXListener。

6.2.3 事件包

对 java.awt 中组件实现事件处理，必须使用 java.awt.event 包，所以在程序开始应加入 import java.awt.event.*语句。Swing 组件不仅能使用 AWT 事件包中的事件类型，而且还有自己的事件包，用于处理 Swing 特有的事件。主要介绍 AWT 事件包中的事件类。

事件类所在的包：java.awt.event 包（如图 6-10 所示），它提供 AWT 事件所需的类和接口。

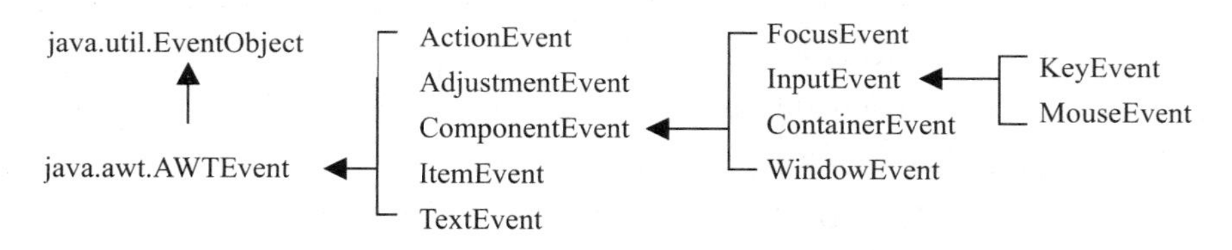

图 6-10 AWT Event 类的相关事件

Java 包中定义的一系列的事件、监听器接口及事件发生时可调用的方法如表 6-1 所示。

表 6-1 AWT 事件的主要监听器接口

事件类别	描述信息	接口名	方法
ActionEvent	激活组件	ActionListener	actionPerformed(ActionEvent e)
ItemEvent	选择某些项目	ItemLisener	itemStateChanged(ItemEvent)
MouseEvent	鼠标单击等	MouseListener	mousePressed(MouseEvent) mouseReleased(MouseEvent) mouseEntered(MouseEvent) mouseExited(MouseEvent) mouseClicked(MouseEvent)
	鼠标移动	MouseMotionListerner	mouseDragged(MouseEvent) mouseMoved(MouseEvent)

续表

事件类别	描述信息	接口名	方法
KeyEvent	键盘输入	KeyListener	KeyPressed(KeyEvent) KeyReleased(KeyEvent) KeyTyped(KeyEvent)
WindowEvent	窗口收到窗口级事件	WindowListener	windowClosing(WindowEvent) windowClosed(WindowEvent) windowOpened(WindowEvent) windowIconified(WindowEvent) windowDeiconified(WindowEvent) windowActivated(WindowEvent) windowDeactivated(WindowEvent)
TextEvent	文本发生改变	TextListener	textValueChanged(TextEvent)

6.2.4 事件的主要处理方法

- 选择组件作为事件源，不同类型的组件会产生特定类型的事件。
- 定义要接收并处理某种类型的事件 XxxEvent，注册相应的事件监听器类。通过调用组件方法：eventSourceObject.addXxxListener(XxxListener)方法向组件注册事件监听器。
- 实现 XxxListener 类的实例对象，据此可作为事件的监听器对象。注册与注销监听器常可表述成为如下两种形式：
  - ✓ 注册监听器:public void add<ListenerType> (<ListenerType>listener);
  - ✓ 注销监听器:public void remove<ListenerType> (<ListenerType>listener)。
- 事件源通过实例化事件类激发并产生事件，事件将被传送给已注册的一个或多个监听器。事件监听器在接收到激发事件信号后负责实现相应的事件处理方法。

6.2.5 键盘事件

当用户使用键盘上一个键进行操作时，就导致这个组件触发 KeyEvent 事件，监听器要完成对事件的响应，就要实现 KeyListener 接口。

程序中要添加如下语句：import java.awt.event.*。并实现 KeyListener 接口，将键盘注册给监听器：eventSourceObject.addKeyListener(listener)。

用 KeyEvent 类的 getKeyCode()方法可以判断哪个键被按下、点击或释放并获取其键码值，getKeyChar()方法被按下键的字符。

【例 6-6】键盘激活事件的处理示例。

```
import java.awt.*;
import java.awt.event.*;
import javax.swing.*;
public class KeyPressTest
{
    public static void main(String[] args)
    {
        KeyFrame frame = new KeyFrame();
        frame.setDefaultCloseOperation(JFrame.EXIT_ON_CLOSE);
        frame.show();
    }
}
```

```
class KeyFrame extends JFrame
{
    public KeyFrame()
    {   JLabel jl=new JLabel();
        jl.setText("请在键盘上按下任意一个键，观察变化！");
        setTitle("KeyPress");
        setSize(280, 190);
        KeyPanel panel = new KeyPanel();    // 将panel加入到frame
        panel.add(jl);
        add(panel);
    }
}
class KeyPanel extends JPanel implements KeyListener
{
    public char KeyChar = ' ';
    public int KeyInputCode = -1;
    public String KeyText = "";
    public boolean isChar = false;
    public KeyPanel()
    {
        addKeyListener(this);                 //注册监听器
    }
    public boolean isFocusTraversable()     //允许面板获得焦点
    {
        return true;
    }
    public void paintComponent(Graphics g)
    {
        super.paintComponent(g);
        if(KeyInputCode == -1)
            g.drawString("输入键码：",90, 70);    //输出输入键码
        else
            g.drawString("输入键码："+ KeyInputCode,90, 70);

        //输出键的名称
        g.drawString("键码名称："+ KeyText, 90, 90);
        //输出字符
        g.drawString("对应字符："+ KeyChar, 90, 110);
    }
    public void keyPressed(KeyEvent event)
    {
        KeyInputCode = event.getKeyCode();//获取输入键码
    }
    public void keyReleased(KeyEvent event)
    {
        KeyInputCode = event.getKeyCode();
        KeyText = event.getKeyText(KeyInputCode);  //获取键的名称

        //获取字符
```

```
            if(!isChar)
            {
                KeyChar= ' ';
            }
            isChar = false;
            repaint();
        }
        public void keyTyped(KeyEvent event)
        {
            KeyChar = event.getKeyChar();
            isChar = true;
        }
    }
```

运行效果如图 6-11 所示。

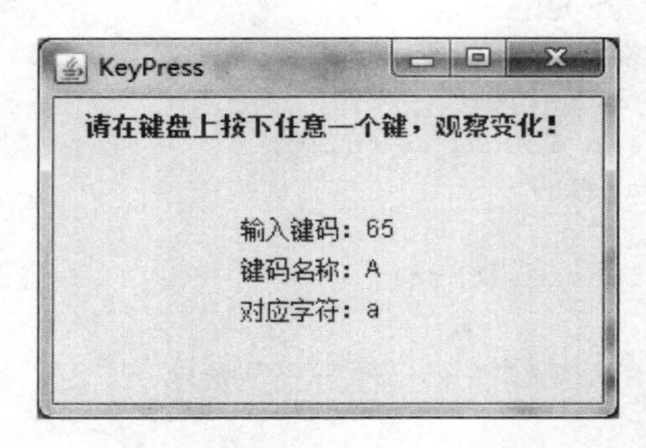

图 6-11　键盘事件示例效果图

6.2.6　鼠标事件

当鼠标键被按下、释放、单击、移动、拖动时会产生 MouseEvent 鼠标事件，为响应并处理该事件，可通过 java.awt.evten 包中的 MouseListener、MouseMotionListener 实现接口。鼠标事件的类型是 MouseEvent，组件触发鼠标事件时 MouseEvent 类自动创建一个事件对象。

Java 使用 MouseListener 与 MouseMotionListener 两接口来处理鼠标事件。

1．MouseListener 接口

MouseListener 接口事件源使用 addMouseListener (MouseListener Listener)方法获取监视器，用户可通过 4 种操作使得事件源触发鼠标事件。

2．MouseMotionListerner 接口

MouseMotionListerner 接口事件源使用方法 addMouseMotionListener 和（MouseMotion List-ener Listener）获取监视器，用户通过 2 种操作使得事件源触发鼠标事件。

当处理鼠标事件时，程序经常关心鼠标在当前组件坐标系中的位置，以及触发鼠标事件使用的是鼠标的左键或右键等信息。

MouseEvent 类中常用方法如下。

Point getPoint（）：取得鼠标按键按下时的坐标，并以 Point 类的对象返回。

int　getX（）：取得鼠标按键按下时的 x 坐标。

int　getY（）：取得鼠标按键按下时的 y 坐标。

【例 6-7】鼠标事件的处理，演示鼠标在各种操作下程序做出的反应。

```
import java.awt.Container;
import java.awt.event.MouseEvent;
import java.awt.event.MouseListener;
import javax.swing.JFrame;
import javax.swing.JLabel;
/**
 * 监听鼠标事件
 * 可以看出，当双击鼠标时，第一次的点击会触发一次单击事件
 */
public class MouseEventTest extends JFrame {
    public MouseEventTest() {

        Container container = getContentPane();
        container.addMouseListener(new MouseListener() {
            @Override
            public void mouseClicked(MouseEvent e) {
                // TODO Auto-generated method stub
                System.out.print("单击了鼠标按键，");
                int i = e.getButton();
                if (i == MouseEvent.BUTTON1)
                    System.out.print("单击的是鼠标左键，");
                if (i == MouseEvent.BUTTON2)
                    System.out.print("单击的是鼠标中键，");
                if (i == MouseEvent.BUTTON3)
                    System.out.print("单击的是鼠标右键，");
                int clickCount = e.getClickCount();
                System.out.println("单击次数为" + clickCount + "下");
            }
            @Override
            public void mousePressed(MouseEvent e) {
                // TODO Auto-generated method stub
                System.out.print("鼠标按键被按下，");
                int i = e.getButton();
                if (i == MouseEvent.BUTTON1)
                    System.out.println("按下的是鼠标左键");
                if (i == MouseEvent.BUTTON2)
                    System.out.println("按下的是鼠标中键");
                if (i == MouseEvent.BUTTON3)
                    System.out.println("按下的是鼠标右键");
            }
            @Override
            public void mouseReleased(MouseEvent e) {
                // TODO Auto-generated method stub
                System.out.print("鼠标按键被释放，");
                int i = e.getButton();
                if (i == MouseEvent.BUTTON1)
```

```
            System.out.println("释放的是鼠标左键");
        if (i == MouseEvent.BUTTON2)
            System.out.println("释放的是鼠标中键");
        if (i == MouseEvent.BUTTON3)
            System.out.println("释放的是鼠标右键");
    }
    @Override
    public void mouseEntered(MouseEvent e) {
        // TODO Auto-generated method stub
        System.out.println("光标移入组件");
    }
    @Override
    public void mouseExited(MouseEvent e) {
        // TODO Auto-generated method stub
        System.out.println("光标移出组件");
    }
  });
}
public static void main(String[] args) {
    // TODO Auto-generated method stub
 MouseEventTest frame = new MouseEventTest();
    frame.setTitle("鼠标事件测试");
    JLabel jl=new JLabel("移动鼠标进来试试");
    frame.add(jl);
    frame.setVisible(true);
    frame.setDefaultCloseOperation(JFrame.EXIT_ON_CLOSE);
    frame.setBounds(0, 0, 300, 100);
}
}
```

运行效果如图 6-12 所示。

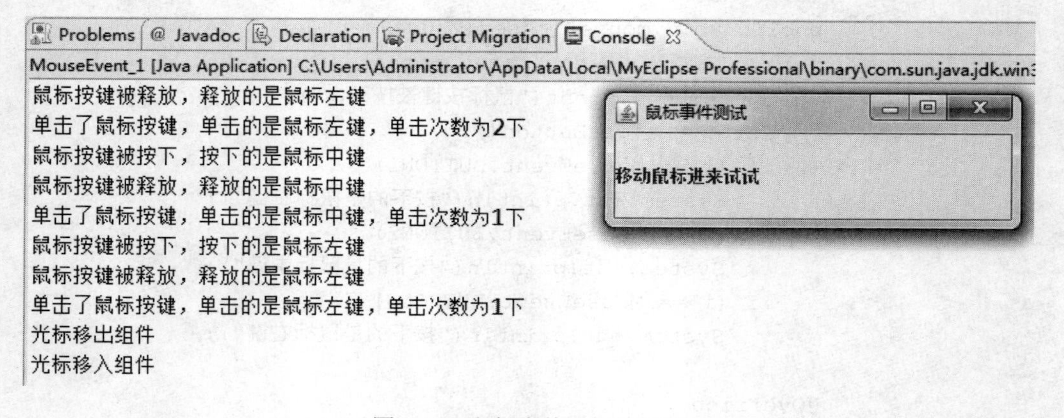

图 6-12　鼠标事件示例效果图

✧ 编码实施

具有了事件处理机制的注册功能：

```
import java.awt.*;
import javax.swing.*;
public class RegPanelEve extends JFrame {
    public static final int WIDTH = 250;
    public static final int HEIGHT = 260;
    public RegPanelEve(){
        setTitle("新用户注册");
        JPanel panel = new JPanel(new FlowLayout(FlowLayout.CENTER));

        JPanel p1 = new JPanel();
        JLabel label1 = new JLabel("用  户  名");
        final JTextField userName = new JTextField(15);
        p1.add(label1);
        p1.add(userName);
        panel.add(p1);

        JPanel p2 = new JPanel();
        JLabel label2 = new JLabel("密        码");
        final JPasswordField password = new JPasswordField(15);
        p2.add(label2);
        p2.add(password);
        panel.add(p2);

        JPanel p3 = new JPanel();
        JLabel label3 = new JLabel("确认密码");
        final JPasswordField rePassword = new JPasswordField(15);
        p3.add(label3);
        p3.add(rePassword);
        panel.add(p3);

        JPanel p4 = new JPanel();
        JLabel sex = new JLabel("性别");
        final ButtonGroup group = new ButtonGroup();
        final JRadioButton male = new JRadioButton("男");
        male.setSelected(true);
        final JRadioButton female = new JRadioButton("女");
        group.add(male);
        group.add(female);
        p4.add(sex);
        p4.add(male);
        p4.add(female);
        panel.add(p4);

        JPanel p5 = new JPanel();
        JLabel label5 = new JLabel("感兴趣的领域");
        final JCheckBox love1 = new JCheckBox("运动");
        final JCheckBox love2 = new JCheckBox("旅游");
```

```
        final JCheckBox love3 = new JCheckBox("美食");
        p5.add(label5);
        p5.add(love1);
        p5.add(love2);
        p5.add(love3);
        panel.add(p5);

        JPanel p6 = new JPanel(new FlowLayout(FlowLayout.CENTER));
        JButton reg = new JButton("注册");
        JButton clear = new JButton("清空");
        p6.add(reg);
        p6.add(clear);
        panel.add(p6);

        reg.addActionListener(new ActionListener() {
            @Override
            public void actionPerformed(ActionEvent e) {
                String un = userName.getText();
                char[] pw = password.getPassword();
                char[] repw = rePassword.getPassword();
                String                              sex                              =
male.isSelected()?male.getText():female.getText();
                StringBuffer hobby = new StringBuffer();
                hobby.append(love1.isSelected()?love1.getText()+"\t":"");
                hobby.append(love2.isSelected()?love2.getText()+"\t":"");
                hobby.append(love3.isSelected()?love3.getText()+"\t":"");
               System.out.println("新用户加入到数据库中");
                System.out.println("用户名: "+un);
                System.out.println("密码: " + new String(pw));
                System.out.println("重复密码: " + new String(repw));
                System.out.println("性别: " + sex);
                System.out.println("兴趣: "+hobby);
            }
        });

        clear.addActionListener(new ActionListener() {
            @Override
            public void actionPerformed(ActionEvent e) {
                userName.setText("");
                password.setText("");
                rePassword.setText("");
                male.setSelected(true);
                love1.setSelected(false);
                love2.setSelected(false);
                love3.setSelected(false);
            }
        });
```

```
            this.add(panel);
            init();
        }
        public void init(){
            Dimension d = Toolkit.getDefaultToolkit().getScreenSize();
            int x = (int) ((d.getWidth()-WIDTH)/2);
            int y = (int) ((d.getHeight()-HEIGHT)/2);
            setBounds(x, y, WIDTH, HEIGHT);
            setDefaultCloseOperation(EXIT_ON_CLOSE);
            setVisible(true);
            setResizable(false);
        }
        public static void main(String[] args) {
            new RegPanelEve();
        }
    }
```

✧ 调试运行

（1）编写一个类 RegPanelEve，创建一个窗口。

（2）创建各种组件对象，并在窗口添加按钮组件。

（3）为按钮添加鼠标事件监听器。

（4）运行程序生成窗体，用鼠标操作窗口中的按钮，观察控制台的输出。

（5）如果程序需要处理某种事件，就需要实现响应的事件监听接口。用专门的顶层类来实现监听接口，优点是可以使处理事件的代码与创建 GUI 界面的代码分离，缺点是在监听类中无法直接访问组件。在监听类的事件处理方法中不能直接访问事件源，而必须通过事件类的 getSource（）方法来获得事件源。

（6）还可以用容器类实现某个监听接口的方法实现事件监听。由于 Java 支持一个类实现多个接口，因此容器类可以实现多个监听接口。容器中的组件将容器实例本身注册为监听器。

✧ 维护升级

如果实现一个监听接口，必须实现接口中所有的方法，否则这个类必须声明为抽象类。为简化编程，Java 针对大多数事件监听器接口定义了相应的已经实现了接口功能的实现类：事件适配器。引入事件适配器的宗旨是使监听器的创建变得更加简便。

在适配器类中系统自动实现相应监听器接口中所有的方法（只写出空的方法体），但不做任何事情。编程时定义继承事件适配器类的监听器，只重写需要的方法。

如 Java 在提供 WindowListener 接口的同时，又提供了 WindowAdapter 类，该类实现了 WindowListener 类接口。因此，可使用 WindowAdapte 的子类创建的对象作为监视器，在子类中重写所需要的接口方法即可。

常用的事件适配器类如下。

- KeyAdapter：键盘事件适配器。
- MouseAdapter：鼠标事件适配器。

- MouseMotionAdapter：鼠标运行事件适配器。
- WindowAdapter：窗口事件适配器。
- Focus Adapter：焦点事件适配器。
- ComponentAdapter：组件事件适配器。
- ContainerAdapter：容器事件适配器。

【例 6-8】用鼠标事件适配器实现在窗口中用鼠标拖动画线的例子。

```
import javax.swing.*;
import java.awt.*;
import java.awt.event.*;
public class DrawLine1 extends JApplet
{   int x1,y1,x2,y2;
    public void init()
    { addMouseListener(new M1());                   //参数定义为 M1 类的对象
        addMouseMotionListener(new M2());           //参数定义为 M2 类的对象
    }
    public void paint(Graphics g)
    {   g.drawLine(x1,y1,x2,y2);
        }
    class M1 extends MouseAdapter
    //创建 MouseAdapter 的子类 M1
    {   public void mousePressed(MouseEvent e)
        {   x1=e.getX();
            y1=e.getY();
        }
    }
    class M2 extends MouseMotionAdapter
    //创建 MouseMotionAdapter 的子类 M2
    {   public void mouseDragged(MouseEvent e)
        {   x2=e.getX();
            y2=e.getY();
            repaint();
        }
    }
}
```

对于定义的事件监听器接口，它可能包含多个事件处理函数，但在使用时往往只需要关注其中的某一种事件，此时就可以使用抽象化适配来将事件监听器接口进行抽象化，给出每一个接口函数的默认实现，这样在需要使用时只需要重写自己需要的函数即可。我们可能只需要处理其中的某一个事件，却不得不编写所有的接口函数，此时就可以使用适配器类了。该适配器实现了和鼠标相关的事件接口，并为这些接口中的所有函数都提供了默认的实现，这些实现的代码为空，表示什么也不做。通过使用事件的适配器，可以让我们的代码只关注自己的事件，而不必造成不必要的代码浪费。这就是适配器模式在事件处理中的应用。

任务3 实现员工信息系统主界面

✧ 需求分析

设计员工信息系统主界面，设计主要包括系统登录、用户管理、系统管理和关于等操作。

1. 需求描述

本任务主要介绍在图形界面设计中常用的一些组件创建与设置，如按钮、文本框、文本域、单选按钮、复选框、列表框等的使用。

2. 运行结果

在Eclipse中运行该项目，结果如图6-13、图6-14所示。

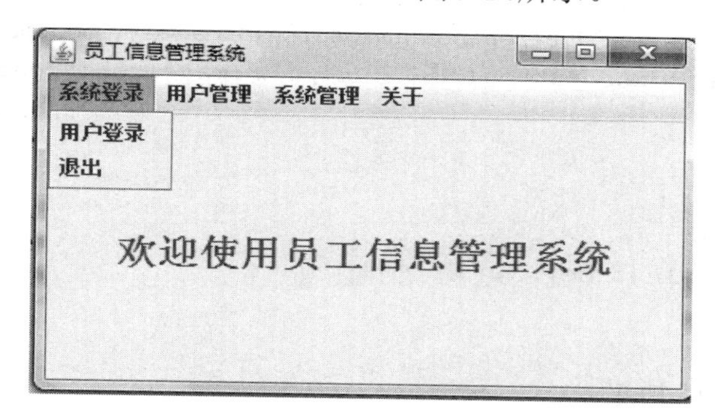

图 6-13　用户注册窗口

图 6-14　用户登录窗口

✧ 知识准备

6.3.1　按钮

按钮是图形界面设计中最常见的一个组件，按钮常用于执行某种动作，用户可通过按钮来触发一个事件。按钮 JButton 类是 AbstractButton 类(javax.swing 包)的子类，按钮是 JButton 类的对象。下面是该类的构造方法和实例方法。

按钮类 JButton 的构造方法：

```
JButton()                              //创建按钮
JButton(String c)                   //创建带指定文本的按钮
JButton(Icon image)              //创建带指定图标的按钮
JButton(String c ,Icon image)    //创建带指定文本和图标的按钮
```

【例 6-9】设计按钮举例。

```
import javax.swing.*;
import java.awt.*;
public class exa_9
{
 public static void main(String [] args)
 {
  JFrame f=new JFrame("带图标的按钮示例");
  f.setLocation(10,100);
  f.setSize(300,200);
  f.setLayout(new FlowLayout());
  f.setDefaultCloseOperation(JFrame.EXIT_ON_CLOSE);      //设置关闭窗体
  JButton ja = new JButton("确定");                      //带文本的按钮
  JButton jb = new JButton(new ImageIcon("images\\logo.png"));   //带图像的按钮
  JButton jc = new JButton("图标按钮",new ImageIcon("images\\logo2.png"));
                                                       //带文本和图像的按钮
  f.add(ja);
  f.add(jb);
  f.add(jc);
  f.setVisible(true);
 }
}
```

运行效果如图 6-15 所示。

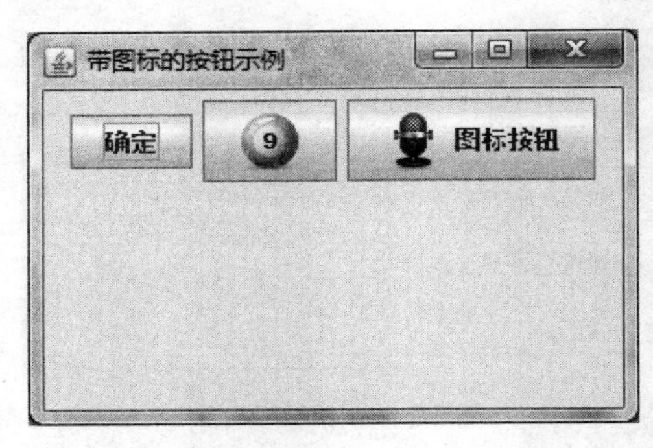

图 6-15　按钮示例效果

【程序分析】此程序创建了三种不同风格的按钮。

6.3.2　文本框、文本域和标签

1．文本框（JTextField）

文本框是一种接收用户键盘输入和修改文本的组件（JTextField），可以处理单行文本输入。

文本框 JTextField 的构造方法：

```
JTextField()                          //创建空白文本的文本框
JTextField(int columns)               //创建指定宽度的空白文本域
JTextField(String text)               // 创建指定文本的文本框
JTextField(String text, int columns)  //指定宽度并初始化文本
```

文本框其他常用方法：

public void setText(String c)：设置显示文本

public String getText()：取得显示文本

public void setFont(Font f)：字体设置

public void setForeground(Color c)：前景色设置

public void　setEditable(boolean b)：是否可编辑

public　StringgetSelectedText()：获得选择的文本

public void setEchoChar(char c)：设置回显字符

另外常用一种用于输入密码的文本框类 JPasswordField 是文本框的一个子类,通常要用它的 setEchoChar(char)方法设置回显字符。

例如：
```
JPasswordField userpwd = new JPasswordField();
userpwd.setEchoChar( '*' );
```

【例 6-10】文本框应用举例。

```
import javax.swing.*;
import java.awt.*;
public class exa_10
{
 public static void main(String [] args)
 {
  JFrame f=new JFrame("JTextField小例子");
  f.setLocation(10,100);
  f.setSize(600,400);
  f.getContentPane().setLayout(new FlowLayout());
  JTextField tf = new JTextField("请输入");
  tf.setColumns(10);          //设置显示的列数
  f.getContentPane().add(tf);
   JPasswordField pf = new JPasswordField("France");
  pf.setColumns(10);          //设置显示的列数
  f.getContentPane().add(pf);
  f.setVisible(true);
  f.pack();                   //使得窗体自动调整大小，以正好显示一个文本框
 }
}
```

运行效果如图 6-16 所示。

图 6-16　JTextField 示例效果

2．文本域（JTextArea）

文本域是可以完成多行输入的可编辑文本框（JTextArea）。

文本域 JTextArea 的构造方法：

```
JTextArea()                              //创建一个新的文本域组件
```

```
JTextArea(int rows,int columns)      //创建指定行数 r 列数 c
JTextArea(String c)                  //创建指定初始化文本
JTextArea(String c,int rows, int columns) //创建指定初始化文本及文本的行数和列数
如:
JTextArea  jta=new JTextArea("请选择",5, 1)
```

常用设置方法:

public void setLineWrap(boolean wrap):设置是否自动换行

public void setRows(int rows):设置行数

public void setColumns(int columns):设置列数

public void append(String c):将指定文本追加到该组件结尾

public void insert(String c ,int i):将指定文本插入文本区的指定行

public void setWrapStyleWord(Boolean b):设置换行方式

说明:颜色、字体、内容的设置同上(setForeground、setFont、setText).

文本域是没有滚动条的,如果需要可以自行添加一个滚动面板,JScrollPane 就是为其他组件提供滚动条的组件,滚动面板提供了一种不能作为自由窗口的通用容器。它应当总是和一个容器相关联(例如,框架),可以将一个面板加入到滚动面板中,配置面板的布局管理器,并在那个面板中放置组件。

创建方法如下:

```
JScrollPane jsp=new JScrollPane ( jta);
```

这样就创建了一个带滚动条的文本域,当文本超出它的显示范围时,滚动条才显示出来。

3.标签(JLabel)

用来显示文字、图标(还可以文字与图标同时显示)。

标签 JLabel 的构造方法:

JLabel( ):创建一个空标签

JLabel(String c):创建指定文本的标签

JLabel(String c ,Int i):创建指定文本的标签,并且指定对齐方式(取值为 LEFT 、CENTER、 RIGHT)

JLabel(String c ,Icon image,Int i):创建指定文本、图标和对齐方式的标签

标签常用设置方法:

设置标签显示文本:public void setText(String c)

取得标签显示文本:public String getText()

字体设置:public void setFont(Font f)

前景色设置:public void setForeground(Color c)

【例 6-11】标签和文本区应用举例。

```
package test;

import javax.swing.*;
import java.awt.*;
public class exa_11
{
 public static void main(String [] args)
 {
```

```
    JFrame f=new JFrame("标签和文本区");
    f.setLocation(10,100);
    f.setSize(600,400);
    f.setDefaultCloseOperation(JFrame.EXIT_ON_CLOSE);  //设置关闭窗体
    f.getContentPane().setLayout(new FlowLayout());
    ImageIcon ii = new ImageIcon("images\\logo.jpg");
    JLabel jl = new JLabel("带图标的标签", ii, JLabel.CENTER);
    f.getContentPane().add(jl);
    JTextArea jta=new JTextArea("本学期开设：",5,10);
    jta.setLineWrap(true);
    jta.append("\n计算机基础\njava程序设计\n数据结构");
    f.getContentPane().add(jta);
    f.pack();
    f.setVisible(true);
  }
}
```

运行效果如图6-17所示。

图6-17 标签和文本区示例效果

6.3.3 复选框与单选按钮

复选框和单选按钮都是从JToggleButon类继承而来的，而JToggleButon类和按钮类又是AbstractButton的子类，所以按钮、单选按钮和复选框在使用上有很多相似之处。如它们都有选中和未选中状态，可以通过isSelected()方法的返回值获得，true表示处于选中状态，false表示未选中状态。

1．复选框

复选框(JCheckBox)是用来做多项选择的，一般都与文本标签一起出现，当选中时包含一个复选标记，否则没有。

复选框JCheckBox的构造函数：

JCheckBox(String c)：创建带有文本标签的复选框

JCheckBox(String c ,Boolean state)：创建带文本和初始状态的复选框

JCheckBox(Icon image)：创建带图标的复选框

JCheckBox(Icon image,Boolean state)：创建带图标和初始状态的复选框
JCheckBox(String c ,Icon image)：创建带文本和图标的复选框
JCheckBox(String c ,Icon image ,Boolean state)：创建带文本、图标和初始状态的复选框
JcheckBoxo 类的其他函数如下：
void setText(String text)：设置按钮的文本
String getText()：返回按钮的文本
void addItemListener(ItemListener i)：添加监听器，监听选择框 ItemEvent 事件
boolean getState()：返回选择框的状态，若选中则返回 true，否则返回 false

【例 6-12】复选框编程方法举例。

```
import javax.swing.*;
import java.awt.event.*;
public class exa_12  extends JFrame implements ItemListener
{
private JPanel panel=new JPanel();
private JCheckBox cb1,cb2,cb3;//声明三个复选框
private JTextField txtField=new JTextField(30);//声明并创建文本框
private String favor="";
public exa_12()
{
setTitle("JCheckBox");
setSize(320,120);
setDefaultCloseOperation(JFrame.EXIT_ON_CLOSE);
getContentPane().setLayout(new BorderLayout());
cb1=new JCheckBox("唐诗");         //创建复选框 music
cb2=new JCheckBox("宋词");         //创建复选框 reading
cb3=new JCheckBox("散文");         //创建复选框 basketball
panel.add(lb);
panel.add(cb1);                    //在面板中添加三个复选框
panel.add(cb2);
panel.add(cb3);
   //在框架中添加文本框
cb1.addItemListener(this);
cb2.addItemListener(this);
cb3.addItemListener(this);
getContentPane().add(panel,"North");            //在框架中,添加面板
getContentPane().add(txtField,"South");
}
public void itemStateChanged(ItemEvent e)  //选择复选框后进行的处理方法
{
if(e.getSource()==cb1)
{
if(music.isSelected()==true)
favor+="爱好唐诗";
}
else if(e.getSource()==cb2)
{
```

```
if(reading.isSelected()==true)
favor+="喜欢宋词";
}
else if(e.getSource()==cb3)
{
if(basketball.isSelected()==true)
favor+="钟情散文";
}
txtField.setText(favor);
}
public static void main(String[]args)
{
JFrame frame=new exa_12();
frame.show();
}
}
```

运行效果如图 6-18 所示。

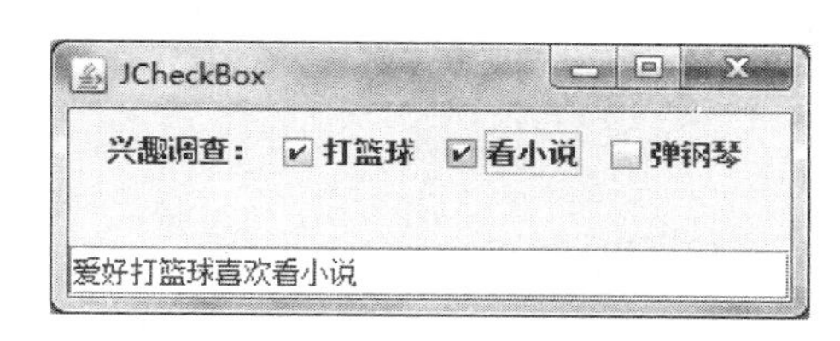

图 6-18　复选框示例效果

【程序分析】选取或不选取(取消)一个复选框的事件将被送往 ItemListener 接口。所传递的 ItemEvent 包含 getSource()方法，它根据实际情况返回事件源，还有另外一个方法是 isSelect()，返回复选框的选中状态是什么，根据事件源的状态作相应的处理。如果想给复选框添加图片，可以用 setSelectedIcon()方法设置，setRolloverIcon()方法的功能为设置鼠标经过复选框上方时显示的图片。

2．单选按钮

单选按钮（JRadioButton）是在选中的时候包含一个点的圆圈，否则是空的。单选按钮必须成组出现，而且一组中只能有一个被选中，因此需要一个专门的类 ButtonGroup 来管理。添加到 ButtonGroup 的多个单选按钮中，如果有一个被选中，同组中的其他单选按钮都会自动改变其状态为未选择状态。

JRadioButton 类具有和 JCheckBox 相同参数和功能的构造函数。

和 JCheckBox 共同的方法有：

setSelected(Boolean)：设置选中状态

isSelected()：返回是否选中状态

为了对复选框进行分组，也可以使用 ButtonGroup 进行分类，创建 ButtonGroup 类的对象的示例：

```
ButtonGroup bg = new ButtonGroup();
JRadioButton radio1 = new JRadioButton("红",true);
bg.add(radio1);
```

```
    radio1.addActionListener(this);
    JRadioButton radio2 = new JRadioButton("绿",false);
    bg.add(radio2);
        radio2.addActionListener(this);
```

注意初始化单选按钮时，一组中只能有一个状态设为 true。当用户单击单选按钮时既能产生一个动作事件（ActionEvent），也能产生一个选项事件（ItemEvent），因此创建单选按钮时既可以注册动作监听器还可以注册选项监听器。

【例 6-13】单选按钮编程方法举例。

```
import javax.swing.*;
import java.awt.*;
import java.awt.event.*;
public class exa_13  extends JFrame implements ActionListener
{
  JFrame frm=new JFrame("Single selection.");
  JRadioButton rb1,rb2;            //声明单选按单选按钮
  JPanel p1;
  ButtonGroup bgroup;              //声明单选按钮组
  JLabel label;
public exa_13()
{
  setForeground(Color.white);
  setSize(200,100);
  setDefaultCloseOperation(JFrame.EXIT_ON_CLOSE);
  frm.getContentPane().setLayout(new BorderLayout());   //边界布局管理器
  bgroup=new ButtonGroup();
  label=new JLabel("鱼和熊掌不能兼得，请选择");
  getContentPane().add(label,"North");
  p1=new JPanel();
  rb1=new JRadioButton("鱼",true);
  rb1.addActionListener(this);       //给单选按钮注册动作监听器
  bgroup.add(rb1);                   //在选择框组中创建单选按钮
  p1.add(rb1);                       //把单选按钮添加到一个面板中
  rb2=new JRadioButton("熊爪",false);
  rb2.addActionListener(this);
  bgroup.add(rb2);
  p1.add(rb2);
  getContentPane().add(p1,"South");//把面板添加到窗体中
}

public void actionPerformed(ActionEvent e)//选中单选按钮后的处理方法式
 {
  if(e.getSource()==rb1)
   {    label.setText("您选择了鱼");
   }
  if(e.getSource()==rb2)
   {    label.setText("您选择了熊爪");
   }
```

```
        }
      public static void main(String[]args)
        {
         JFrame frame=new exa_13();
         frame.show();
        }
      }
```

程序运行结果如图 6-19 所示。

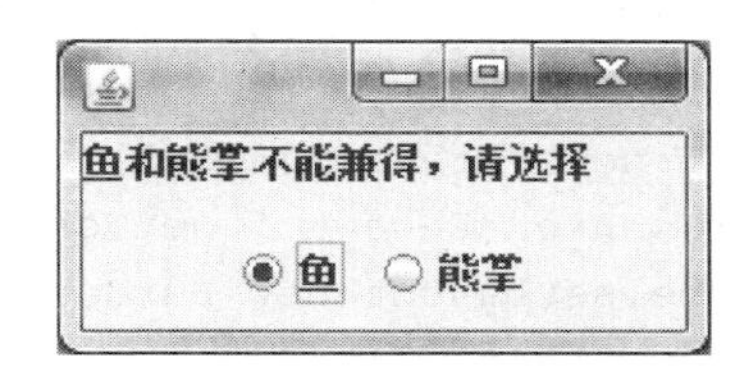

图 6-19　单选按钮示例效果

【程序分析】程序中创建了两个单选按钮 rb1 和 rb2，初始化对象 rb1 的选中状态为 true，选中单选按钮后的处理方式是通过 getSource（）方法获得事件源，按事件源的不同作出相应的处理。

6.3.4　列表框和组合框

1．列表框（JList）

列表框（JList）在可供选择的选项很多时，可向用户呈现一系列的选项来供选择，用户可以从中选择一个或多个条目，既可显示字符串，也可显示图标。

列表框 JList 的构造函数：

public JList( ) ：创建一个新的列表框。

public JList(ListModel dataModel)：创建一个列表框，用它显示指定模型中的元素。

public JList (Object [] listData)：创建一个列表框，以显示指定数组 listData 的元素。

JList 不支持滚动。要启用滚动，可使用下列代码：

```
JScrollPane myScrollPane=new JScrollPane();
    myScrollPane.getViewport().setView(dataList);
```

建立列表框前，需要指定所显示的列表条目：

```
String[] season = {"Spring", "Summer ", "Autumn", "Winter " };
JList seasonList = new JList(season);
```

如果列表框比较长而显示屏幕比较小，可以设置列表中显示的行数，同时使用滚动条：

```
nList.setVisibleRowCount(4);
tScrollPane = new JScrollPane(seasonList);
```

框就会产生选择事件。将列表选择监听器 ListSelection-

Changed 方法。

号需要使用 getSelectedValues 方法和

```
t.getSelectedValues();
.getSelectedIndices();
```

【例 6-1

```
package test;
import java.awt.*;
import java.awt.event.*;
import javax.swing.*;
import javax.swing.event.*;
    public class JListDemo implements ListSelectionListener {
        JList list = null;
        JLabel label = null;
        String[] s = { "美国", "中国", "英国", "法国", "意大利", "澳洲", "韩国" };
        public JListDemo() {
            JFrame f = new JFrame("JList");
            Container contentPane = f.getContentPane();
            contentPane.setLayout(new BorderLayout());
            label = new JLabel();
            list = new JList(s);
            list.setVisibleRowCount(5); // 设定列表方框的可见栏数
            list.setBorder(BorderFactory
                    .createTitledBorder("您最喜欢到哪个国家玩呢? "));
            list.addListSelectionListener(this);
            contentPane.add(label, BorderLayout.NORTH);
            // 给列表方框添加滚动栏
            contentPane.add(new JScrollPane(list), BorderLayout.CENTER);
            f.pack();
            f.show();
            f.addWindowListener(new WindowAdapter() {
                public void windowClosing(WindowEvent e) {
                    System.exit(0);
                }
            });
        }
        public void valueChanged(ListSelectionEvent e) {
            int tmp = 0;
            String stmp = "您目前选取: ";
            //利用JList类所提供的getSelectedIndices()方法可得到用户所选取的所有项目
            int[] index = list.getSelectedIndices();
            //index值，这些index值由一个int array返回
            for (int i = 0; i < index.length; i++)
            {
                tmp = index[i];
                stmp = stmp + s[tmp] + "  ";
            }
            label.setText(stmp);
        }
        public static void main(String args[]) {
            new JListDemo();
        }
    }
```

运行结果如图 6-20 所示。

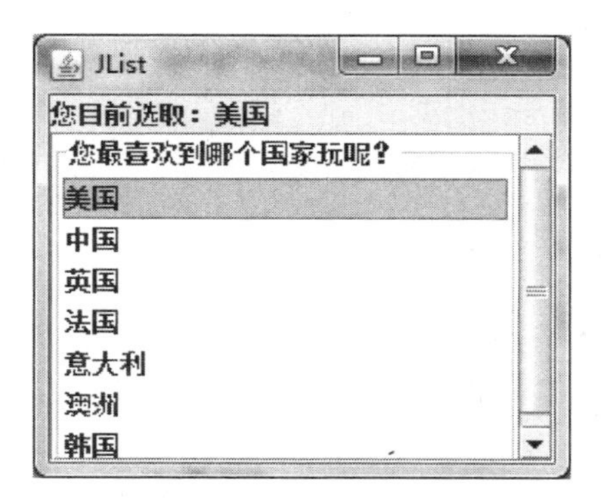

图 6-20　列表框示例效果

2．组合框（JComboBox）

组合框与列表框相似，不同的是下拉列表框只能从列表中选择一个选项，当用户单击旁边下拉箭头按钮时，选项列表才打开，可以较少地占用图形化用户界面的空间。组合框完成了下拉列表的功能之外，还可以输入参数。当 JComboBox 的 setEditable（）方法设置为 true 的时候，用户可以在文本域中输入文本。

JComboBox 具有与 JList 相同参数和功能的构造函数。

JComboBox 的相关方法：

```
addItem(int)               //向组合框内添加选项
    getItemAt(int)         //返回 int 指定的索引位置的列表项目的文本（从 0 开始）。
    getItemCount()         //返回列表中的项目的数量
    getSelectedIndex()     //返回列表中的当前选择项目的索引位置
    getSelectedItem()      //返回当前选择项目的文本
    getSelectedIndex(int)          //选择指定索引位置的项
    getSelectedIndex(Object)       //选择列表中指定的对象
    setMaximumRowCount(int)        //设置在组合框中的一次显示的行的数量
```

组合框主要的两种事件是 ActionEvent 和 ItemEvent,当需要注册一个动作监听器 ActionListener 时，要用 actionPerformed（ActionEvent e）方法处理事件，当需要注册选项监听器 ItemListener 时，要用 itemStateChanged(ItemEvent e)方法处理事件。

【例 6-15】组合框的使用。

```
package test;

import java.awt.*;
import java.awt.event.*;
import javax.swing.*;
public class Example6_4{
    public static void main(String args[]){
        ComboBoxDemo mycomboBoxGUI = new ComboBoxDemo();
    }
}
class ComboBoxDemo extends JFrame implements ActionListener,ItemListener{
    public static final int Width = 350;
```

```
    public static final int Height = 150;
    String proList[] = { "踢足球","打篮球","打排球" };
    JTextField text;
    JComboBox comboBox;
    public ComboBoxDemo(){
        setSize(Width,Height);
        setTitle("组合框使用示意程序");
        Container conPane = getContentPane();
        conPane.setBackground(Color.BLUE);
        conPane.setLayout(new FlowLayout());
        comboBox = new JComboBox(proList);
        comboBox.addActionListener(this);
        comboBox.addItemListener(this);
        comboBox.setEditable(true);//响应键盘输入
        conPane.add(comboBox);
        text = new JTextField(10);
        conPane.add(text);
        this.setVisible(true);
    }
    public void actionPerformed(ActionEvent e){
        if(e.getSource()==comboBox)
            text.setText(comboBox.getSelectedItem().toString());
    }
    public void itemStateChanged(ItemEvent e){
        if(e.getSource()==comboBox){
            text.setText(comboBox.getSelectedItem().toString());
        }
    }
}
```

运行效果如图6-21所示。

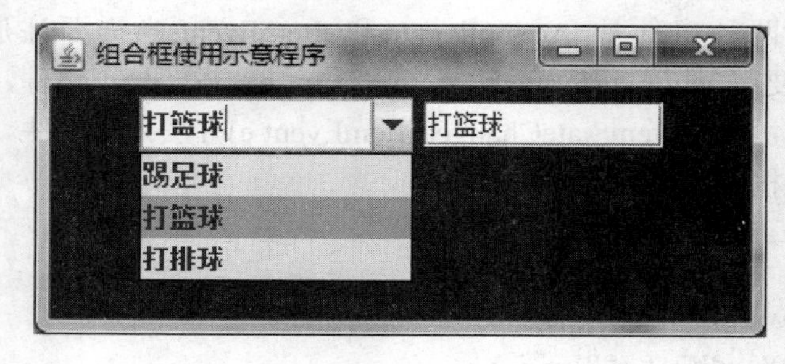

图6-21　组合框示例效果

6.3.5 对话框

对话框分为模式对话框和无模式对话框。

模式对话框必须在用户处理完后才允许用户与主窗口继续进行交互。无模式对话框允许用户同时在对话框和程序剩余部分中输入信息。

Java Swing 中创建对话框使用 JOptionPane 和 JDialog 两个类，前者用来创建标准对话框，后者用于创建用户自定义的对话框。标准对话框都是模态对话框，后者 JDialog 创建的对话框既可是模态的也可是非模态的对话框。

1．JOptionPane 对话框类

在 Swing 中，JOptionPane 提供了四种简单的对话框。

消息对话框（Message Dialog）：是在对话框上显示出一段信息，目的是告知用户一些相关信息，因此 Message Dialog 只会有一个确定按钮，让用户看完信息后就可以关闭这个对话框。

确认对话框（Confirm Dialog）：目的是让用户对某个问题选择“Yes”或“No”，,可算是一种相当简单的是非选择对话框。

选项对话框（Option Dialog）：显示选择对话框，这类对话框可以让用户自定义对话类型，最大的好处是可以改变按钮上的文字。

输入对话框（Input Dialog）：显示一条消息并等待用户的输入。可以让用户输入相关信息,当用户按下确定钮后,系统会得到用户所输入的信息，输入对话框不仅可以让用户自输入文字,也可以显示出 ComboBox 组件让用户选择相关信息，避免用户输入错误，当用户输入完毕按下确定按钮时会返回用户输入的信息，若按下取消则返回 null 值。

这四个标准对话框分别用于提出问题，警告用户，提供简要的重要消息的小窗口。

Confirm Dialog 目的是让用户对某个问题选择"Yes"或"No",可算是一种相当简单的是非选择对话框。

ConfirmDialog 的构造方法如下：

```
    int showConfirmDialog(Component parentComponent,Object message)
    int showConfirmDialog(Component parentComponent,Object message,String
title,int optionType)
    int showConfirmDialog(Component parentComponent,Object message,String
title,int optionType, int messageType)
    int showConfirmDialog(Component parentComponent,Object message,String
title,int optionType, int messageType,con icon)
```

同样的，其他三个对话框的几种创建方法也都类似，就不再一一列出。

说明：

显示确认对话框,这类对话框通常会问用户一个问题,然后用户回答是或不是,例如当我们修改了某个文件的内容却没存盘就要离开时，系统大部分都会跳出确认对话框。询问我们是否要存储修改过的内容，确认对话框方法有 6 种参数。

parentComponent：是指产生对话框的组件为何，通常是指 Frame 或 Dialog 组件。

message：是指要显示的组件，通常是 String 或 Label 类型。

title：对话框标题列上显示的文字。

icon：若你不喜欢 java 给的图标,你可以自己自定图标。

messageType：指定信息类型,共有 5 种类型,分别是：ERROR_MESSAGE，INFORMATION_MESSAGE，WARING_MESSAGE，QUESTION_MESSAGE，PLAIN_MESSAGE（不显示图标）。指定类型后对话框就会出现相对应的图标。

optionType：底部按钮的类型和对话框有密切的关系，对于 showMessageDialog 和 showInputDialog 对话框来说，只能有一组标准按钮，分别是 OK 和 OK/CANCEL。对于 showConfirmDialog，按钮可以包括四种：DEFAULT_OPTION，YES_NO_OPTION，YES_NO_

CANCEL_OPTION，OK_CANCEL__OPTION。

以 showConfirmDialog 为例：

```
int selection = JOptionPane.showConfirmDialog(
    DialogFrame.this,                      //父窗口
    "Are you sure?", "Logout",             //消息以及对话框标题
    JOptionPane.OK_CANCEL_OPTION,          //底部按钮类型
    JOptionPane.WARNING_MESSAGE);          //消息类型
```

这四个对话框可以采用 showXXXDialog()来显示，如 showConfirmDialog()显示确认对话框、showInputDialog()显示输入文本对话框、showMessageDialog()显示信息对话框、show OptionDialog()显示选择性的对话框。

使用 JOptionPane 对象所得到的对话框是 modal 为 true 形式,也就是说必须先关闭对话框窗口才能回到产生对话框的母窗口上。

2．JDialog 对话框类

使用 JOptionPane 能轻易地显示想要的模态对话框，可说相当方便，若这些模式还是无法满足你的需求，就可以使用 JDialog 来自行设计你的对话框。

JDialog 类的构造方法：

JDialog()：建立一个无模态的对话框，没有 title 也不属于任何事件窗口组件。

JDialog(Dialog owner)：建立一个属于 Dialog，non-modal 形式，也没有 title 的对话框。

JDialog(Dialog owner,Boolean modal)：建立一个属于 Dialog 的对话框，可决定 modal 形式，但没有 title。

Dialog(Dialog owner,String title)：建立一个属于 Dialog 组件的对话框，为 non-modal 形式，对话框上有 title。

JDialog(Dialog owner,String title,Boolean modal)：建立一个属于 Dialog 组件的对话框，可决定 modal 形式，且对话框上有 title。

JDialog(Frame owner)：建立一个属于 Frame 组件的对话框，为 non-modal 形式，也没有 title。

JDialog(Frame owner,Boolean modal)：建立一个属于 Frame 组件的对话框，可决定 modal 形式，但没有 title。

JDialog(Frame owner,String title)：:建立一个属于 Frame 组件的对话框，为 non-modal 形式，对话框上有 title。

JDialog(Frame owner,String title,Boolean modal)：建立一个属于 Frame 组件的对话框，可决定 modal 形式，且对话框上有 title。

下面看一个对话框的实例。

【例 6-16】对话框的使用举例。

```
import java.awt.*;
import javax.swing.*;
import java.awt.event.*;
public class exa_18 implements ActionListener
{
    JFrame f = null;
    JLabel label = null;
    public exa_16()
    {   f = new JFrame("OptionPane Demo");
```

```
        Container contentPane = f.getContentPane();
        JPanel panel = new JPanel();
        panel.setLayout(new GridLayout(2,2));

        JButton b = new JButton("Show DEFAULT_OPTION");
        b.addActionListener(this);
        panel.add(b);
        b = new JButton("Show YES_NO_OPTION");
        b.addActionListener(this);
        panel.add(b);
        b = new JButton("Show YES_NO_CANCEL_OPTION");
        b.addActionListener(this);
        panel.add(b);
        b = new JButton("Show OK_CANCEL_OPTION");
        b.addActionListener(this);
        panel.add(b);
        label = new JLabel(" ",JLabel.CENTER);
        contentPane.add(label,BorderLayout.NORTH);
        contentPane.add(panel,BorderLayout.CENTER);
        f.pack();
        f.setVisible(true);
        f.addWindowListener(new WindowAdapter() {
            public void windowClosing(WindowEvent e) {
                System.exit(0);
            }
        });
    }
public static void main(String[] args)
    {
        new exa_16();
    }

    public void actionPerformed(ActionEvent e)
    {
  //处理用户按钮事件,默认的 messageType 是 JoptionPane.INFORMATION_MESSAGE.
        String cmd = e.getActionCommand();
        String title = "Confirm Dialog";
        String message ="";
        int messageType = JOptionPane.INFORMATION_MESSAGE;
        int optionType = JOptionPane.YES_NO_OPTION;

        if(cmd.equals("Show DEFAULT_OPTION")) {
            optionType = JOptionPane.DEFAULT_OPTION;
            message = "Show DEFAULT_OPTION Buttons";
        } else if(cmd.equals("Show YES_NO_OPTION")) {
            optionType = JOptionPane.YES_NO_OPTION;
            message = "Show YES_NO_OPTION Buttons";
```

```
        } else if(cmd.equals("Show YES_NO_CANCEL_OPTION")) {
            optionType = JOptionPane.YES_NO_CANCEL_OPTION;
            message = "Show YES_NO_CANCEL_OPTION Buttons";
        } else if(cmd.equals("Show OK_CANCEL_OPTION")) {
            optionType = JOptionPane.OK_CANCEL_OPTION;
            message = "Show OK_CANCEL_OPTION Buttons";
        }

        int result = JOptionPane.showConfirmDialog(f, message,
                    title, optionType, messageType);

        if (result == JOptionPane.YES_OPTION)
            label.setText("您选择: Yes or OK");
        if (result == JOptionPane.NO_OPTION)
            label.setText("您选择: No");
        if (result == JOptionPane.CANCEL_OPTION)
            label.setText("您选择: Cancel");
        if (result == JOptionPane.CLOSED_OPTION)
            label.setText("您没做任何选择，并关闭了对话框");
    }
}
```

当点击 Show YES_NO_OPTION 后弹出如图 6-22 所示的对话框。

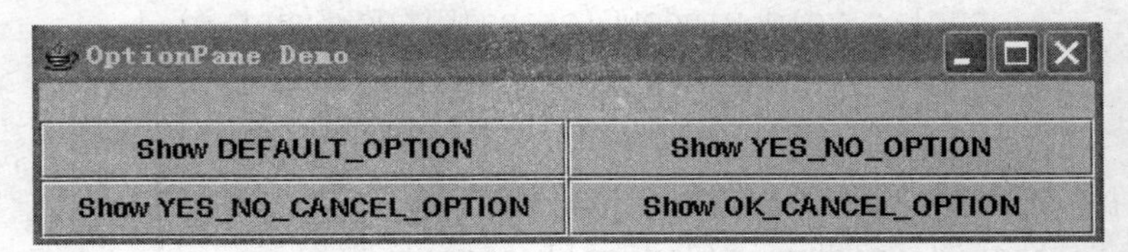

图 6-22 对话框示例运行效果

6.3.6 菜单

在 GUI 界面中，菜单一般位于窗口上方标题栏下面的位置，它是一个图形用户界面不可缺少的组成部分。

一个菜单由三部分组成：

JmenuBar——菜单栏，在其中可以加入菜单（JMenu）。

JMenu——菜单，其中可加入菜单项（JMenuItem）和菜单（Jmenu）。

JMenuItem——菜单项，直接指向一个具体的操作。

创建菜单的四个步骤：

- 创建菜单栏、菜单以及子菜单，创建菜单项；
- 将菜单项加入到子菜单或菜单中；
- 将子菜单加入到菜单中；
- 将菜单加入到菜单栏中。

菜单、菜单项和菜单栏的构造方法：

菜单栏——JMenuBar()

菜单——JMenu(String s)

菜单项——JMenuItem(String s)

1．菜单条（JMenuBar）

一个菜单条组件是一个水平菜单。菜单条只能加入到一个框架中，一个菜单条由多个菜单（JMenu）组成，每个JMenu在JMenubar中都表示为字符串。在一个时刻，一个框架中可以显示一个菜单条，菜单条不支持监听者。

```
JFrame f = new JFrame("JMenuBar");
JMenuBar mb = new JMenuBar();
f.setJMenuBar(mb);
```

2．菜单（JMenu）

提供了一个基本的下拉式菜单，它可以加入到一个菜单条或者另一个菜单中。它在菜单栏下以文本字符串形式显示，而在用户单击它时，则以弹出式菜单显示。

```
JMenuBar mb = new JMenuBar();
JMenu m1 = new JMenu("File");
JMenu m2 = new JMenu("Edit");
JMenu m3 = new JMenu("Help");
mb.add(m1);
mb.add(m2);
mb.setHelpMenu(m3);
f.setMenuBar(mb);
```

3．菜单项（JMenuItem）

菜单项组件通常被加入到菜单中，以构成一个完整的菜单，以文本字符串形式显示，可以具有图标。JMenuItem 的外观可以修改，如字体、颜色、背景、边框等。除字符串外，在JMenuItem 中还可以添加图标。通常将一个 ActionListener 加入到一个菜单项对象中，以提供菜单的行为。

```
JMenu m1 = new JMenu("File");
JMenuItem mi1 = newJ MenuItem("New");
JMenuItem mi2 = new JMenuItem("Load");
JMenuItem mi3 = new JMenuItem ("Save");
JMenuItem mi4 = new JMenuItem("Quit");
mi1.addActionListener(this);
m1.add(mi1);
mi2.addActionListener(this);
m1.add(mi2);
mi3.addActionListener(this);
m1.addSeparator();
m1.add(mi3);
```

4．复选菜单项（JCheckboxMenuItem）

一个可复选的菜单项，可以在菜单上有选项(“开”或“关”) 。应当用 ItemListener 接口来监视复选菜单。当复选框状态发生改变时，就会调用 itemStateChanged()方法。

```
JMenu m1 = new JMenu("File");
JMenuItem mi1 = new JMenuItem("Save");
JCheckboxMenuItem mi2 = new JCheckboxMenuItem("checkbox1");
mi1.addItemListener(this);
```

```
mi2.addItemListener(this);
m1.add(mi1);
m1.add(mi2);
```

5．单选按钮菜单项（JRadioButtonMenuItem）

属于一组菜单项中的一个菜单项,该组中只能选择一个项。被选择的项显示其选择状态。选择此项的同时,其他任何以前被选择的项都切换到未选择状态。要控制一组单选按钮菜单项的选择状态，应使用 JButtonGroup 对象。

```
JButtonGroup group = new JButtonGroup();
JRadioButtonMenuItem mi3 = new JRadioButtonMenuItem("Forward");
group.add(mi3);
m1.add(mi3);
```

6．弹出式菜单（JPopupMenu）

弹出式菜单提供了一种独立的菜单，它可以在任何组件上显示。你可以将菜单项和菜单加入到弹出式菜单中去。

```
JFrame f = new JFrame("PopupMenu");
JButton b = new JButton("Press Me");
JPopupMenu p = new JPopupMenu("Popup");
JMenuItem s = new JMenuItem("Save");
JMenuItem l = new JMenuItem("Load");
b.addActionListener(this);
f.add(b,Border.Layout.CENTER);
p.add(s);
p.add(l);
f.add(p);
```

为了显示弹出式菜单，必须调用显示方法。显示需要一个组件的引用，作为 x 和 y 坐标轴的起点

```
public void actionPerformed(ActionEvent ev) {
     p.show(b, 10, 10);    }
```

7．菜单项的使用状态

菜单项能够被选取取决于菜单项的启用和禁用状态：

```
mi3.setEnabled(false); //初始设置"Save"菜单项为禁用状态
```

8．快捷键和加速器

快捷键显示为带有下画线的字母，加速器则显示为菜单项旁边的组合键。

```
m1.setMnemonic('F');    //设置菜单的快捷键
JMenuItem exitItem = new JMenuItem("Exit", 'T'); //设置"Exit"菜单项的快捷
                                                  键为"T"。
exitItem.setAccelerator(KeyStroke.getKeyStroke(     KeyEvent.VK_T,
InputEvent.CTRL_MASK));     //设置"Exit"菜单项的加速器为"Ctrl+T"。
```

【例 6-17】菜单使用举例。

```
import javax.swing.*;
import java.awt.*;
import java.awt.event.*;
import java.io.*;
class MenuTest extends JFrame implements ActionListener{
```

```
    JMenuBar mb;
    JMenu file,send,config;
    JMenuItem op,ne,ex,sf,se,fc,bc;
    JTextArea t;
    MenuTest(){
        super("我的第一个菜单");
        this.setDefaultCloseOperation(3);
        menuInit();
        this.setJMenuBar(mb);
        this.getContentPane().add(new JScrollPane(t));
        this.setBounds(300,200,600,400);
        this.setVisible(true);
    }
    void menuInit(){
        mb=new JMenuBar();//创建各个菜单项
        file=new JMenu("文件");
        send=new JMenu("发送");
        op=new JMenuItem("打开");
        ne=new JMenuItem("新建");
        ex=new JMenuItem("退出");
        se=new JMenuItem("邮件");
        sf=new JMenuItem("保存");
        mb.add(file);//把 file 菜单加入菜单栏
        file.add(op);//把各菜单项放到 file 菜单里
        file.add(ne);       file.add(send);
        file.addSeparator();
        file.add(ex);       send.add(se);
        send.add(sf);
        config=new JMenu("设置");
        fc=new JMenuItem("前景色");
        bc=new JMenuItem("背景色");
        mb.add(config);
        config.add(fc);
        config.add(bc);
        //给各个菜单项注册事件监听器
        op.addActionListener(this);
        ne.addActionListener(this);
        sf.addActionListener(this);
        ex.addActionListener(this);
        fc.addActionListener(this);
        bc.addActionListener(this);
        //设置文本区
        t=new JTextArea();
        t.setLineWrap(true);
    }
```

```
    //下面是响应各个菜单的事件类
    public void actionPerformed(ActionEvent e){
        Object o=e.getSource();
        JFileChooser f=new JFileChooser();
        JColorChooser cc=new JColorChooser();
        if(o==ne){
            t.setText("");
        }
        else if(o==op){
            f.showOpenDialog(this);      //从文件选择对话框中选择一个文件打开
            try{
                StringBuffer s=new StringBuffer();
                //从指定的文件中读取一个序列化的对象
                FileReader in=new FileReader(f.getSelectedFile());
                while(true){
                    int b=in.read();
                    if(b==-1)break;
                    s.append((char)b);
                }
                t.setText(s.toString());
                in.close();
            }
            catch(Exception ee){}
        }
        else if(o==sf){     //如果选中"保存"则调用保存对话框写入
            f.showSaveDialog(this);
            try{
                FileWriter out=new FileWriter(f.getSelectedFile());
                out.write(t.getText());
                out.close();
            }catch(Exception ee){}
        }
        else if(o==ex)System.exit(0);   //选中"退出"时的响应
        else if(o==bc){
            Color c=cc.showDialog(this,"Please select",Color.white);
            t.setBackground(c);          //根据选择的颜色设置背景色
        }
        else{
            Color c=cc.showDialog(this,"Please select",Color.black);
            t.setForeground(c);          //根据选择的颜色设置前景色
        }
    }
    public static void main(String[] aa){
        new MenuTest(); }
}
```

运行效果如图 6-23 所示。

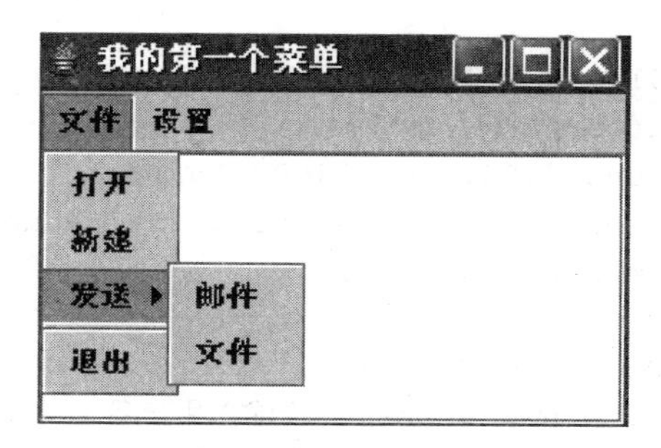

图 6-23　菜单示例运行效果

✧ 编码实施

员工信息系统的主界面实现代码如下：

```
import java.awt.*;
import java.awt.event.*;
import javax.swing.*;

public class EmployerMange{
 JFrame frame = new JFrame();
 Container content = frame.getContentPane();
 JMenuBar menubar = new JMenuBar();
 JMenu LoginMenu = new JMenu("系统登录");
 JMenu UserMangeMenu = new JMenu("用户管理");
 JMenu SchoolMangeMenu = new JMenu("系统管理");
 JMenu HelpMenu = new JMenu("关于");
 public EmployerMange(){
  JMenuItem userLoginMenu= new JMenuItem("用户登录");
  userLoginMenu.addActionListener(new LoginActionListener());
  JMenuItem exitLoginMenu= new JMenuItem("退出");
  LoginMenu.add(userLoginMenu);
  LoginMenu.add(exitLoginMenu);
  menubar.add(LoginMenu);
  menubar.add(UserMangeMenu);
  menubar.add(SchoolMangeMenu);
  menubar.add(HelpMenu);
  frame.setTitle("员工信息管理系统");
  content.add(menubar,BorderLayout.NORTH);
  content.add(new JLabel("<html><font size='6' color='red'>欢迎使用员工信
息管理系统</font></html>",JLabel.CENTER),BorderLayout.CENTER);
//  content.add(new JLabel("欢迎使用员工信息管理系统",JLabel.CENTER),
BorderLayout.CENTER);
  frame.setBounds(450, 200, 400, 400);
  frame.setVisible(true);
 }
 public class LoginActionListener implements ActionListener{
```

```
    public void actionPerformed(ActionEvent e) {

     JFrame frame01 = new JFrame();
     frame01.setTitle("用户登录");
     frame01.setBounds(450, 200, 400, 150);
     Container content01 = frame01.getContentPane();
     JPanel panel = new JPanel();
     JPanel panel01 = new JPanel();
     JPanel panel02 = new JPanel();
     JLabel label01 = new JLabel("请输入用户名: ");
     JLabel label02 = new JLabel("请输入密码:      ");
     JTextField text01 = new JTextField(20);
     JPasswordField text02 = new JPasswordField(20);
     panel01.add(label01);
     panel01.add(text01);
     panel02.add(label02);
     panel02.add(text02);
     panel.add(panel01);
     panel.add(panel02);
     content01.add(panel,BorderLayout.CENTER);
     frame01.pack();
     frame01.setVisible(true);
    }
   }
   public static void main(String[] args) {
    new EmployerMange();
   }
  }
```

✧ 调试运行

1. 编写一个类 EmployerMange，创建一个窗口。
2. 创建各种组件对象，并在窗口添加菜单等组件。
3. 为按钮添加鼠标事件监听器。
4. 运行程序生成窗体和登录窗体。

✧ 维护升级

Java 连接数据库的登录功能参考步骤：

```
  public void jButton1_actionPerformed(ActionEvent actionEvent) {
      String jusername=text01.getText().trim();
      char[] s= text02.getPassword();
      String jpassword=new String(s);
  //如果没输用户名或密码,则提示对不起,请输入用户名或密码
          if(jusername.equals("")||jpassword.equals(""))
          {
             JOptionPane.showMessageDialog(this,"对不起,请输入用户名或密
码.","错误!",JOptionPane.ERROR_MESSAGE);          }
```

```
//如果都有数据了就开始连接数据库验证
        else  {
            try  {
    Class.forName("com.microsoft.jdbc.sqlserver.SQLServerDriver");
    conn=DriverManager.getConnection(URL,USER,PASSWORD);//这里就是连接数据库
了
                String sql="select * from employee  where username='"+
                jusername+"'";
                //执行的 sql 语句,在数据库里查找我们输入的用户名
    stmt=conn.createStatement();
     rs=stmt.executeQuery(sql);
                if(rs.next()) //如果存在,就验证密码
                  {
                   if(rs.getString("password").equals(jpassword))
                               //如果密码正确就提示成功,反之提示密码错误
                   {
                               //new  ErsBlocksGame("aaa");
                      System.out.println("dengluchenggong");
                      }                    else{
                      JOptionPane.showMessageDialog(this,"对不起,密码错误,
请重新输入","登陆失败",JOptionPane.ERROR_MESSAGE);
                      }
                }
    //如果没有查找到用户名就提示重新登录
                else{
                   JOptionPane.showMessageDialog(this,"用户名不存在,请重新
输入","错误!",JOptionPane.ERROR_MESSAGE);
                   }
            }catch(ClassNotFoundException ex){//这后面是抛出异常
              ex.printStackTrace();           }
            catch(SQLException
ex){                ex.printStackTrace();           }
            finally{
              try{
                   if(rs!=null) rs.close();
                   if(stmt!=null) stmt.close();
                   if(conn!=null) conn.close();
                   }
                catch(SQLException ex)
                {
                   ex.printStackTrace();
                   }
              }
            }
    }
```

员工信息系统的其他功能，如增加员工，修改、删除和查询员工信息等功能都可以参考以上实现步骤进行完善，结合 GUI 界面设计和数据库连接知识形成完整的信息系统设计。

任务4 嵌入网页上的Applet 程序

✧ 需求分析

Java Applet 程序是一种在网页上运行的小程序。Applet 使用非常广泛，许多特色网页的交互式效果，都可以通过 Applet 程序实现。本任务将介绍 Applet 程序的编写和运行方式以及如何在 Applet 中加载图形图像和声音等多媒体对象。

1．需求描述

制作一个能用鼠标移动图像的 Applet 小程序。

2．运行结果

运行结果如图 6-24 所示。

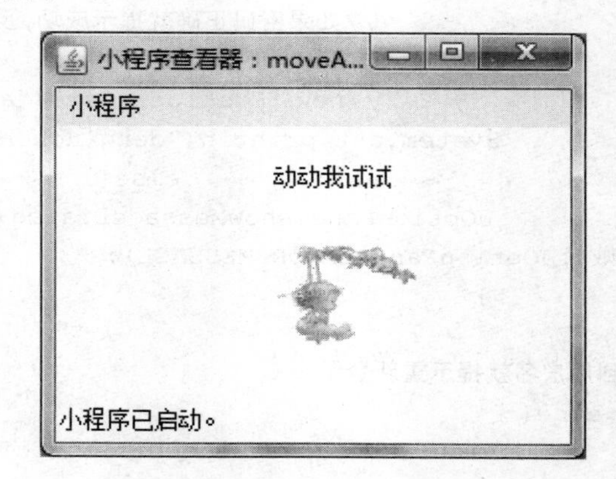

图 6-24 用户注册界面设计

✧ 知识准备

Java 程序分两类：应用程序（Application）和小应用程序（Applet）。

在 Java 语言中，专门为 Applet 提供了一个重要的系统类——Applet 类，该类提供了 Applet 和运行它的环境的接口。Applet 的继承关系如图 6-25 所示。

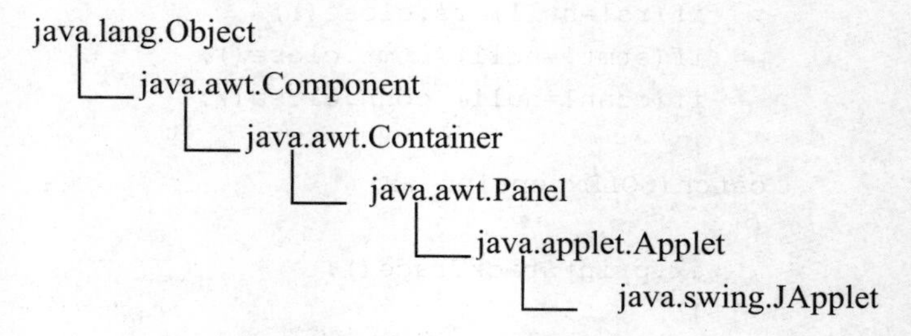

图 6-25 Applet 的继承关系

需要注意的是，在实际应用中经常要用到 javax.Swing 包中的 JApplet 类来编写用于 WWW 的小应用程序。JApplet 类是 Applet 类的直接子类，Swing Applet 的超类，所以如果想使用 Swing 集合来实现 Applet，那么编制的 Applet 应该继承 JApplet 类。

Applet 的安全性也很高，使用 applet 程序完全不必担心黑客入侵和病毒感染，同时由于 applet 程序不能直接访问用户的本地资源，要想实现访问用户的本地资源需要对 Applet 进行授权并进行数字签名。

6.4.1 Applet 类及相关方法

Applet 编程是 Java 编程语言至关重要的独特功能，它不同于一般的 Java 程序的地方是，它能够嵌入在 HTML 网页中，并由支持 Java 的 WEB 浏览器来解释执行，当然我们所使用的 IE 浏览器 3.0 以上的版本是支持 Java 小应用程序的，有了这种小应用程序，我们的网页就能有一定的交互功能。

1．Applet 程序的工作原理

Applet 是一种工作在浏览器上的 java 程序。将编译好的字节码文件保存在特定的 www 服务器上，同一个或另一个服务器上保存着嵌入了该字节码文件名的 html 文件。

当某一个浏览器向服务器请求下载嵌入了 Applet 的 html 文件时，该文件从服务器上下载到客户端，由浏览器解释 html 文件，当遇到<applet>，表明它嵌有一个 Applet，浏览器会根据这个 Applet 的名字和位置自动把字节码从服务器上下载到本地，并利用浏览器本身拥有的 java 解释器执行字节码，这就是 Applet 的工作原理（见图 6-26）。

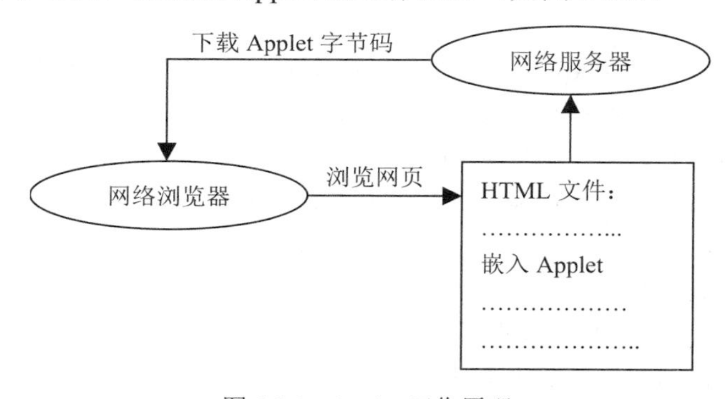

图 6-26　Applet 工作原理

2．Applet 类中的主要方法

Applet 类提供了一些重要的方法，这些方法只能由浏览器在特定的情况下调用，这些重要方法有：init（）、start（）、stop（）和 destroy（）等。通常我们将这四个方法分别对应了 Applet 的初始化、启动、暂停直到消亡的各个阶段。

在实际应用中，用户会根据需要重载这些方法来构造自己需要的 Applet。Applet 程序中的这些方法在不同运行过程中，所处的状态和作用是不同的，它会在适当事件发生时自动调用该实例的几个主要方法：它贯穿整个 Applet 程序的生命中，就是说随着网页被载入，到网页被刷新、切换乃至关闭，各个不同阶段会执行不同的方法，因此不同的方法决定了 Applet 的生命周期，如图 6-27 所示。

（1）public void init（）

该方法用于 Applet 的初始化。当 Applet 被第一次加载时，该方法会被浏览器自动调用，在 Applet 的生命周期中只执行一次。作用是用来通知 Applet 已经被装载到系统中。Applet 的子类可以通过重写本方法使 Applet 完成初始化工作，包括创建所需要的对象、设置初始状态、装载图像或字体、设置参数等。

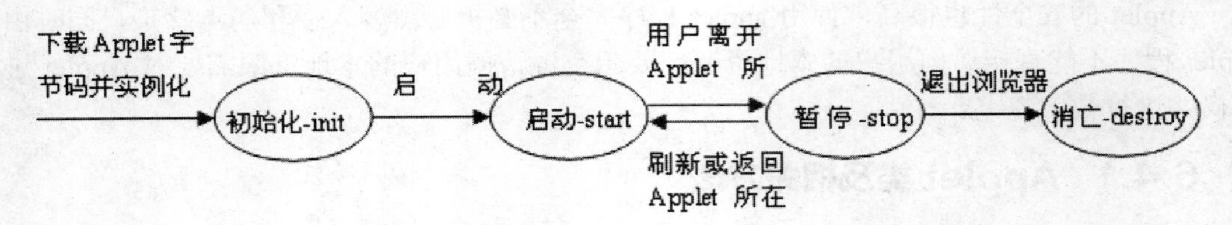

图 6-27 Applet 的生命周期

（2）public void start()

调用 init 方法将小程序初始化后，浏览器将自动调用 start()方法启动运行 Applet 的主线程。当用户刷新包含 Applet 的页面或者从其他页面返回包含 Applet 的页面时，start 方法也会被自动调用。也就是说，start 方法可以被多次调用，这与 init 方法是有区别的。基于这样的原因，可以把只调用一次的代码放在 init 方法中，而不能放在 start 方法中。 Applet 的子类还可以通过重写本方法使 Applet 每次被访问时执行特定的操作。例如，对于一个完成动画功能的 Applet 可以使用 start()方法来重新开始动画演示，使用 stop()方法将动画线程挂起。

小程序被终止后，也可以将它重新启动。在小程序的生命周期中初始化只发生一次，但可以启动多次。

（3）public void stop()

该方法在用户离开包含 Applet 的页面或调用 destroy()方法时会被自动调用，用来通知 Applet 停止执行一些耗费系统资源的活动。和 start 方法一样，stop 方法也可以被多次调用。Applet 的子类可以通过重写本方法使 Applet 每次停止时执行特定的操所。例如，对于一个完成动画功能的 Applet 可以使用 start()方法来重新开始动画演示，使用 stop()方法将动画线程挂起。

终止和启动是成对出现的。用户离开包含小程序的页面或小程序调用 stop()方法时，该小程序将终止。默认情况下，即使在用户离开了页面后，小程序启动的线程仍将继续执行。通过覆盖 stop()方法可以将线程挂起，并在用户返回小程序所在的页面时重新启动它们。

（4）public void destroy()

当用户正常关闭浏览器时，浏览器会调用 destroy 方法来回收系统资源，如回收图形用户界面的系统资源、关闭连接等。在调用本方法前通常先调用 stop()方法。至于 Applet 实例本身，会由浏览器来负责从内存中清除，不需要在 destroy 方法中来清除。

（5）public void paint(Graphics g)

另外一个重要的方法是 paint() 绘图方法，它的作用相当于 Applet 的灵魂，它不是 Applet 类里定义的方法，而是继承自 java.awt.Container 类中的 paint 方法。Applet 本质上是图形方式的，所以尽管你可以提出 System.out.println()的调用请求，通常也不能这样做，你可以通过创建一个 paint()方法在 Applet 的 panel 上绘图。

在下列情况下会使用 paint()方法:

① Applet 被启动之后，自动调用 paint()来重绘。

② Applet 所在的浏览器窗口改变时使用 paint 重画界面。

每当小程序窗口需要重新绘制时，包含小程序的环境（通常是 Web 浏览器）将自动调用 paint()方法；要在小程序中请求重新绘制窗口，可以调用小程序的 repaint()方法，无参数。paint

方法有一个 Graphics 类的对象 g 作为参数,即 paint（Graphics g）{……}。

它用来完成一些较低级的图形用户界面操作，当一个 Applet 类实例被初始化并启动时，浏览器将自动生成一个 Graphics 类的实例 g，并把 g 作为参数传递给 Applet 的 paint 方法，就可以绘制 Applet 的界面。

6.4.2 Applet 程序建立及运行过程

1．源程序的编辑与编译

【例 6-18】先编写 firstApplet.java 文件：

```
import java.applet.*;       //将 java.applet 包中的系统类引入本程序
import java.awt.*;          //将 java.awt 包中的系统类引入本程序
public class firstApplet extends Applet{
    public void paint(Graphics g){
    g.setColor(Color.red);//设置颜色
    g.drawString("我正在学习能嵌入网页的 Applet 小程序!",2,10);
    g.setColor(Color.magenta);
    g.drawString("画条线，画个圆、还能涂色……",2,30);
    g.drawRect(30,45,180,190);            //绘制矩形
    g.drawLine(40,220,200,220);           //绘制直线
    g.drawString("简单吧? ",50,260);       //文字
    g.setColor(Color.blue);
    g.fillRect(40,50,160,150);            //实心矩形
    g.setColor(Color.red);
    g.fillOval(45,55,150,140);            //实心圆
    g.setColor(Color.orange);
    g.fillRect(100,90,40,70);
}
}
```

编译生成 HelloApplet.class 文件。

2．代码嵌入

Applet 中没有 main 方法作为 Java 解释器的入口，必须编写 HTML 文件，把该 Applet 文件嵌入其中。

建立一个网页文件 firstApplet.html，内容如下：

```
<html>
<head><title>我的第一个 JavaApplet 程序</title></head>
<body>
<applet code=" firstApplet.class" width=300 height=200> //这里将显示一个
applet 小程序
</applet>
</body>
</html>
```

注意，这个标记的通用格式与任何其他的 HTML 相同，即，用<和>两个符号来分隔指令。上例中显示的所有部分都是必需的，你必须使用<applet . . .>和</applet>。<applet . . .>部分指明了代码的入口以及宽度和高度，这是一个简单的 html 文件，如果想写出复杂点的程序，可利用其他相关标识实现。

3．Applet 的运行

将该 html 文件放在与 class 文件相同的目录下，用支持 Java 的浏览器如 IE 打开，会看到如图 6-28 效果。

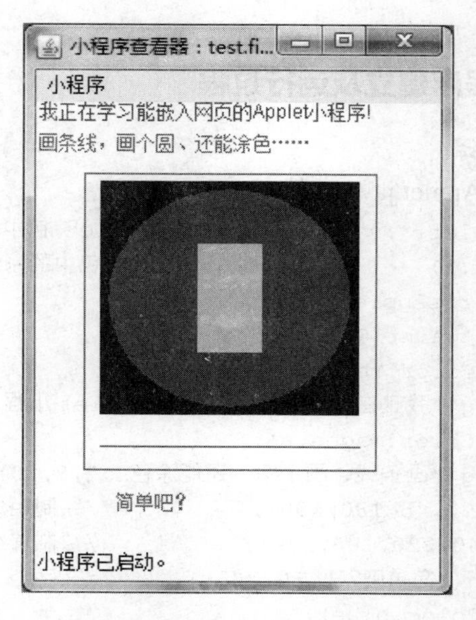

图 6-28　applet 程序示例效果图

6.4.3 Applet 图像技术

Java Applet 常用来显示存储在 GIF 文件中的图像。Java Applet 装载 GIF 图像非常简单，在 Applet 内使用图像文件时需定义 Image 对象。多数 Java Applet 使用的是 GIF 或 JPEG 格式的图像文件。Applet 使用 getImage 方法把图像文件和 Image 对象联系起来。

Graphics 类的 drawImage 方法用来显示 Image 对象。为了提高图像的显示效果，许多 Applet 都采用双缓冲技术：首先把图像装入内存，然后再显示在屏幕上。

Applet 窗口中加载与显示图像的 3 个操作：

- ① 声明 Image 类型的变量；
- ② 使用 getImage()加载图像；
- ③ 使用 drawImage()绘制图像。

首先，装载一幅图像：Java 把图像也当做 Image 对象处理，所以装载图像时需首先定义 Image 对象，然后用 getImage 方法把 Image 对象和图像文件联系起来。格式如下所示：

```
Image picture;
picture=getImage(getCodeBase(),"ImageFileName.GIF");
```

getImage 方法有两个参数。第一个参数是对 getCodeBase 方法的调用，该方法返回 Applet 的 URL 地址，如 www.sun.com/Applet。第二个参数指定从 URL 装入的图像文件名。如果图文件位于 Applet 之下的某个子目录，文件名中则应包括相应的目录路径。

用 getImage 方法把图像装入后，Applet 便可用 Graphics 类的 drawImage 方法显示图像，形式如下所示：

```
g.drawImage(Picture,x,y,this);
```

该 drayImage 方法的参数指明了待显示的图像、图像左上角的 x 坐标和 y 坐标以及 this。第四个参数的目的是指定一个实现 ImageObServer 接口的对象，即定义了 imageUpdate 方法的对象。

✧ 代码实施

```
import java.awt.*;
import java.awt.event.*;
import java.applet.Applet;

public class moveAngelApp extends Applet implements MouseMotionListener,
MouseListener{
    Image img;
    int x=70,y=60,posX=70,posY=60,dx,dy;
    public void init(){
      img=getImage(getCodeBase(),"tianshi.gif");
      addMouseListener(this);
      addMouseMotionListener(this);
    }

    public void mousePressed(MouseEvent e){
      dx=e.getX()-posX;
      dy=e.getY()-posY;
    }
    public void mouseDragged(MouseEvent e){
      x=e.getX()-dx;
      y=e.getY()-dy;
      if(dx>0&&dx<120&&dy>0&&dy<60){
        Graphics g=getGraphics();
        update(g);
      }
    }
    public void paint(Graphics g){
  g.drawString("动动我试试！",110,30);
      g.drawImage(img,x,y,120,60,this);
      posX=x;
      posY=y;
    }
    public void mouseMoved(MouseEvent e){};
    public void mouseReleased(MouseEvent e){};
    public void mouseEntered(MouseEvent e){};
    public void mouseExited(MouseEvent e){};
    public void mouseClicked(MouseEvent e){};
}
```

✧ 调试运行

MyEclipse 里 Applet 程序的运行方法是右键上下菜单有 Applet 运行模式（平时用的 Application 模式）。

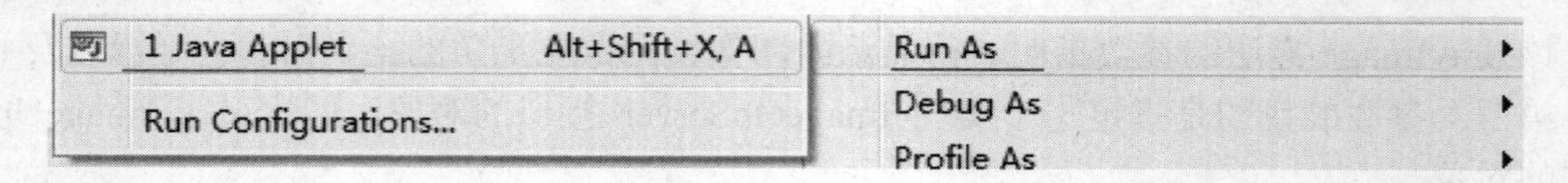

✧ 维护升级

播放声音是 java 对多媒体支持的又一个重要部分，现今流行的声音格式有 wav、mid、au 为扩展名的声音文件，java 支持声音文件的格式相当齐全，主要支持以 au 为名的声音。在 Applet 中专门提供了类 AudioClip 来对声音的支持。在 Applet 中使用 java.applet 类库里的 AudioClip 接口就可以播放声音。

AudioClip 接口定义了 3 个方法：

- loop()——循环播放音乐文件；
- play()——播放音乐文件；
- stop()——停止播放音乐文件。

【例 6-19】演示在 Applet 中播放声音的例子。

```
import java.applet.*;
import java.awt.*;
import java.awt.event.*;
import java.io.File;
import java.net.MalformedURLException;
import java.net.URL;
import javax.swing.*;
public class audioClipApplet extends JApplet{
private AudioClip ac;
private JButton jbtPlay,jbtLoop,jbtStop;
public audioClipApplet()
{
setLayout(new FlowLayout());
jbtPlay=new JButton("播放");
jbtLoop=new JButton("循环");
jbtStop=new JButton("停止");
add(jbtPlay);
add(jbtLoop);
add(jbtStop);
URL urlAudio=this.getClass().getResource("music.wav");
ac=Applet.newAudioClip(urlAudio);
jbtPlay.addActionListener(new ButtonListener());
jbtLoop.addActionListener(new ButtonListener());
jbtStop.addActionListener(new ButtonListener());
}
class ButtonListener implements ActionListener
{
public void actionPerformed(ActionEvent e)
{
if(e.getSource()==jbtPlay)
ac.play();
else if(e.getSource()==jbtLoop)
```

```
ac.loop();
else if(e.getSource()==jbtStop)
ac.stop();
}
}
public static void main(String[] args) throws MalformedURLException
{
JFrame frame = new JFrame("播放音乐的 Applet");
audioClipApplet applate=new audioClipApplet();
frame.add(applate);
frame.setSize(100,100);
JPanel jp=new JPanel();
frame.setDefaultCloseOperation(JFrame.EXIT_ON_CLOSE);
frame.setVisible(true);
}
}
```

运行效果如图 6-29 所示。

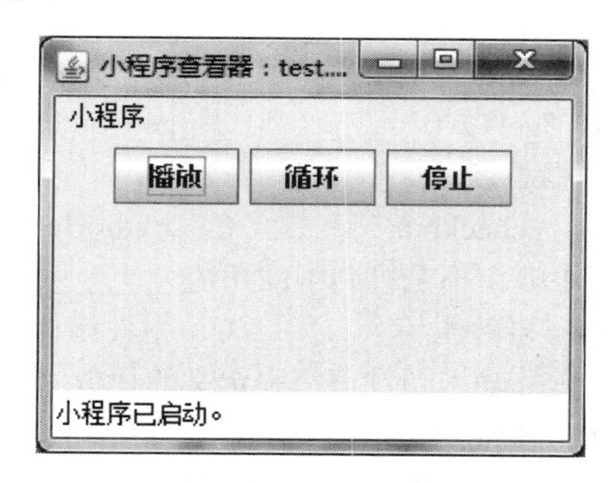

图 6-29 Applet 播放声音的效果

项目实训与练习

一、填空题

1. ________类用于创建一组单选按钮。
2. ________用于安排容器上的 GUI 组件。
3. GUI 是________的缩写。
4. 可以使用________方法，取得当前选择的项目的索引值。
5. 在第一次加载 Applet 时，默认最先执行的方法是________。
6. 用户不能修改的文本称为________。
7. Java 为那些声明了多个方法的 Listener 接口提供了一个对应的________，在该类中实现了对应接口的所有方法。
8. Java 将所有组件可能发生的事件进行分类，具有共同特征的事件被抽象为一个________。
9. addActionListener(this)方法中的 this 参数表示的意思是________________。

10．监听事件和处理事件由________________完成。

11．图形用户界面通过________响应用户和程序的交互，产生事件的组件称为________。

二、简答题

1．举例说明什么是图形用户界面？它的构成成分以及作用有哪些？设计图形用户界面的工作主要有哪两项？

2．简要概述Java语言中事件处理的步骤？事件源是什么？事件监听器是什么？

3．什么是容器的布局策略？常用的布局策略有哪些？

4．什么是Applet？它与Java的应用程序有什么不同？

三、选择题

1．下面组件中不属于swing包的是________。

A．JLabel　　B．Jframe　　C．ButtonGroup　　D．JMenu

2．下面不属于"容器"的是________。

A．文本框　　B．对话框　　C．窗口

3．组件可以被添加到其他组件中去。________

A．正确　　B．不正确

4．要获得一个JLabel上显示的文本信息，应使用________方法。

A．setText　　B．getText　　C．getContent

5．在Java语言中，复选框类名是________。

A．JButton　　B．Jcheckbox　　C．JradioButton　　D．JComboBox

6．在下列事件处理机制中哪个不是机制中的角色________。

A．事件　　B．事件源　　C．事件接口　　D．事件处理者

7．在多行文本框中________输入超过程序中定义的行数。

A．能　　B．不能

8．可以使用________来清除文本框text1中的文本。

A．text1.clearText()　　B．text1.setText("")

C．text1.deleteText()　　D．以上都可以

9．在程序中________，就能够实现多行文本框内容的自动换行。

A．不需要做任何考虑　　B．textarea1.setLineWrap(true)

C．textarea1.setLineWrap(false)　　D．textarea1.noWrap()

10．JFrame默认的布局管理器是下列哪一个________。

A．FlowLayout　　B．BorderLayout

C．GridLayout　　D．CardLayout

四、实训题

1．设计一个合理的加法器（见图6-30）。

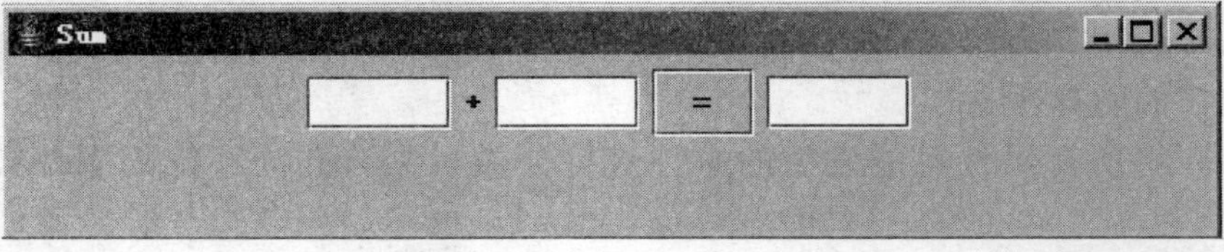

图6-30　运行效果图（一）

2．设计如图 6-31 所示的文本编辑器。

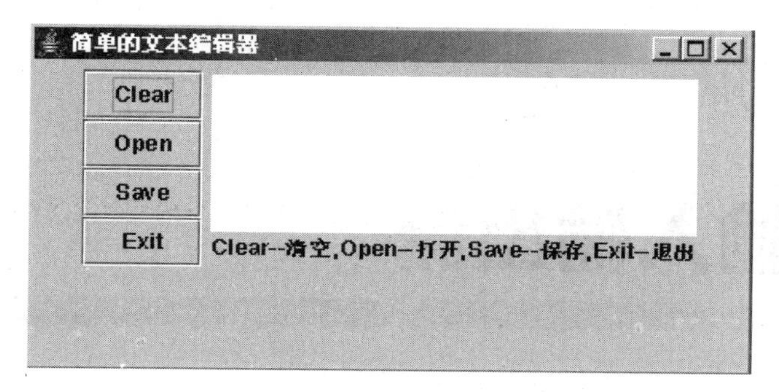

图 6-31 运行效果图（二）

3．编写一个程序如图 6-32 所示，使用户能够使用鼠标在 applet 中绘制一个矩形。按住鼠标左键，确定矩形的左上角，然后拖动鼠标，在需要的位置（即矩形右下角）释放鼠标。另外，在状态栏中显示矩形面积。

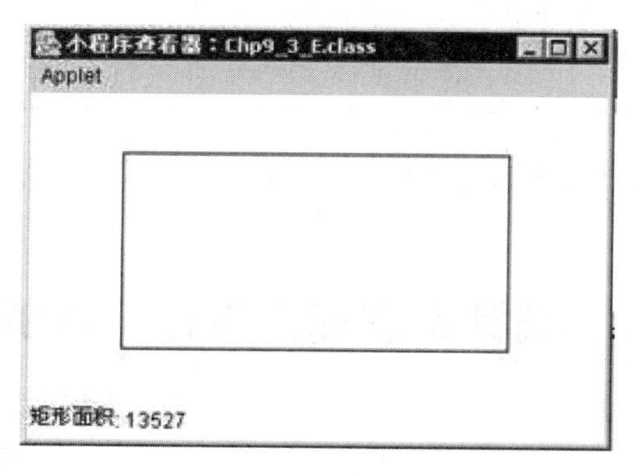

图 6-32 运行效果图（三）

4．编写程序实现如图 6-33 所示界面，实现事件如果按下座位 i 就在控制台中显示“座位 i 被选中”例如按下“座位 0“，则输出座位 0 被选中”。

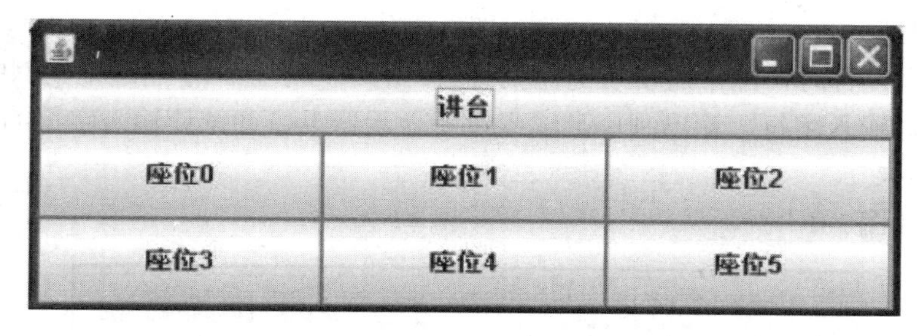

图 6-33 运行效果图（四）

项目 7 输入输出流

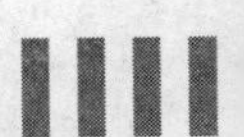

项目目标

本章的主要内容是介绍java中的输入与输出操作。详细介绍java系统中的输入流和输出流的使用。重点对各种字节流和字符流的应用进行了较全面的介绍。通过本章的学习，了解流类的相关层次关系。掌握字节流和字符流在文件读写中的应用方法；熟悉对象序列化的步骤与应用。

项目内容

用 java IO 系统完成对文件读写的任务；利用字节流实现文件的复制过程及对象的序列化。

任务1　统计键盘输入字符个数的程序

✧ 需求分析

在程序开发过程中，经常需要在程序和用户之间进行数据的交互，如将数据写入文件，或从文件中读取数据，有时还需要通过网络进行数据的读写。下面这个案例的功能是统计通过键盘输入字符的个数。

1．需求描述

在程序开发过程时，经常需要从键盘获得用户的输入信息。下面这个案例的功能是从标准输入设备键盘输入字符，输出到标准输出设备显示器上，并统计输入字符的个数。

2．运行结果

运行结果如图 7-1 所示。

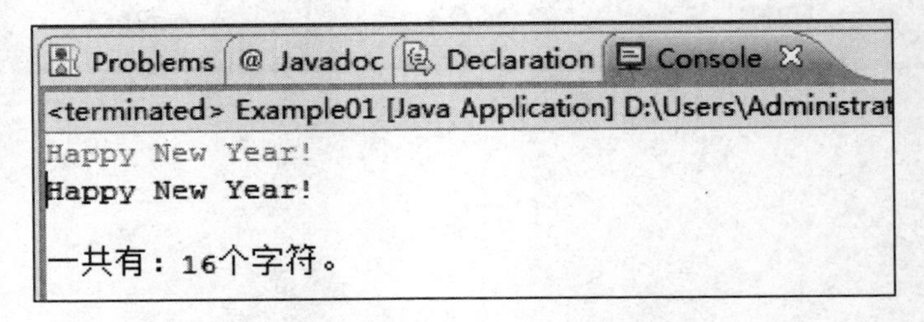

图 7-1　统计字符个数显示

✧ 知识准备

7.1.1 流的概念

什么是“流”呢？可能各位首先想到的是水流，Java io 流和水流是非常类似的。流是一维的，同时流是单向的，对应的操作就是单向读取和单向写入。参照于内存，数据进入内存即为输入，从内存写入其他设备即为输出。

流的方向是重要的，根据流的方向，流可分为两类：输入流和输出流。用户可以从输入流中读取信息，但不能写它。相反，对输出流，只能往输入流写，而不能读它。

流是一个抽象的概念，是一个流动的数据序列，它可以按两个方向输送数据。Java 将流连接到计算机硬件设备，即使它们所连接的硬件设备是不同的，所有的流都以相同的模式工作。例如：一种可以向控制台写入信息的方法也可以向磁盘文件写入信息。流、程序及外围设备之间的关系如图 7-2 所示。

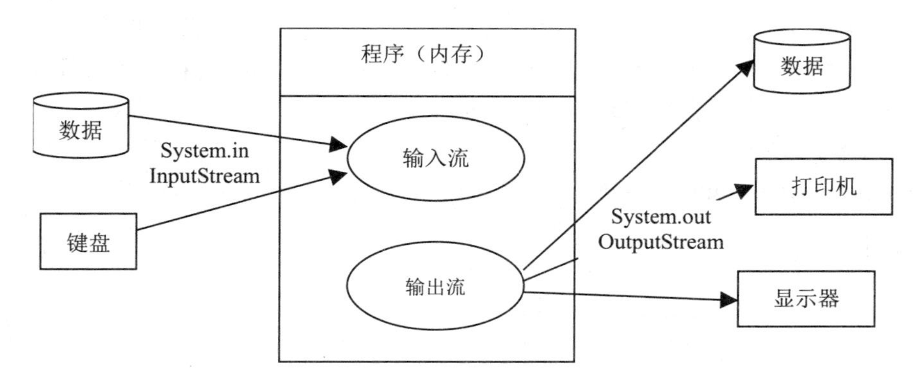

图 7-2　流、程序及外围设备之间的关系

Java 流式 I/O 的工作方式如图 7-2 所示，在图 7-2 中，System.out 是 OutputStream 流的一种，它连接到系统的标准输出（显示器）。程序通过 System.out.println()方法将字符串信息“Hello”写入 OutputStream 流中，然后利用流式传送把信息传递到显示器上，从而完成了输出操作。

相应的，System.in 是 InputStream 流的一种，它连接到系统的标准输入（键盘）。程序通过 System..in.read()方法从 InputStream 流中读取来自键盘的输入，从而完成了输出操作。

按传送的方向，流可以分为输入流和输出流。

（1）输入流将外部数据引入到当前运行的计算机程序中。输入流与数据源相连，程序便可通过输入流从数据源中读取数据。

（2）输出流与外部设备或其他数据接收器相连，程序可向输出流写入数据，将数据送到数据接收器。

对于流可以从不同的角度进行分类，除了上述分为输入流和输出流，还可以按照处理类型的不同分为字节流和字符流，根据流的建立方式和工作原理，可分为节点流和过滤流。

在 Java 开发环境中，主要是由 java.io 包中提供的一系列的类和接口来实现输入/输出处理。所有的 I/O 操作都在 java.io 包之中进行定义，而且整个 java.io 包实际上就是 File、InputStream、OutputStream、Reader、Writer 五个类和一个 Serializable 接口。

7.1.2 流的分类

1．字节流和字符流

较新版本的 Java 系统定义了两种类型的流：字节型和字符型。

字节流提供了处理字节型数据输入和输出的便捷方法。字节流常用于读/写二进制数据，对于文件操作尤其有用。字符流用于处理字符型数据的输入/输出，在某些场合中，字符流比字节流更有效率。

在系统的底层，I/O 操作仍然是字节型的，因而字节流是更基本的流类型。字符流只是为处理字符型数据提供方便和快捷的方法。

2．节点流和处理流

这是 I/O 类的另一种分类方法。节点流是直接连接到数据源（对输入流）或数据目标（对输出量）的一种流。在有 I/O 操作的程序，必须存在节点流，以完成程序对于外部的输入/输出操作。

与节点流不同，处理流不直接与数据源或数据目标相连，而是另外的流进行配合，对数据进行某种处理，完成流的转换或提高读/写流的效率等。处理流不是程序中必须存在的部分，但是如果要想对流进行特定的操作，则需要加入处理流。

3．标准输入/输出

System.in、System.out 和 System.err 是 Java 系统定义的标准输入/输出流。

✧ 编码实施

```
import java.io.IOException;
public class Example01 {

    /**
     * @param args
     * @throws IOException
     */
    public static void main(String[] args) throws IOException {
        // TODO Auto-generated method stub
        int a;
        int count = 0;
        while(( a = System.in.read())!='\n')//按回车键结束程序
        {
            count++ ;
            System.out.print((char)a);
        }
        System.out.println();
        System.out.println("一共有："+count+"个字符。");
    }
}
```

✧ 调试运行

① 执行程序后，用户从键盘输入字符，按回车键后，在显示器上显示输入的内容。按

回车键结束程序时，统计出此次输入的字符个数。

② 在这个例子里，用户输入的是“Happy New Year!”，统计一下应该只有15个字符，为何结果显示是16个字符？这是因为输入结束时用户需要按回车键，程序将回车键按1个字符来统计计算。

✧ 维护升级

如果想让统计结果更容易理解，用户可以修改程序，将count的初始值设为-1即可。所有输入字符统计的程序均存在此问题，读者需注意。

任务2 利用字节流实现文件的复制过程

✧ 需求分析

编写一个程序，可以复制已经存在的文件，并将文件内容输出显示。

1. 需求描述

编写一个程序，将文件readme.txt复制到指定位置，并显示文件内容。

2. 运行结果

运行结果如图7-3所示。

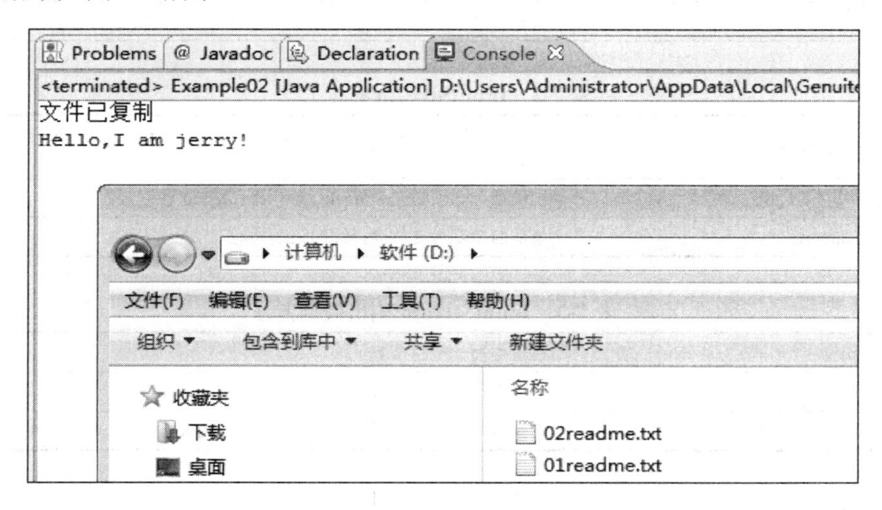

图7-3 文件复制运行结果

✧ 知识准备

在java中，我们可以通过InputStream、OutputStream、Reader、Writer类来处理流的输入与输出。InputStream与OutputStream类通常是用来处理“字节流”，也就是二进制文件。二进制文件是不能被windows中的记事本直接编辑的文件，在读写二进制文件时必须使用字节流，例如word文档、音频和视频文件等。而Reader与Writer类则用来处理“字符流”，也就是纯文本文件，纯文本文件是可以被Windows中的记事本直接编辑的文件。

7.2.1 字节流概述

字节流提供了处理字节的输入/输出方法，它在 Java 中用来处理以字节为主的流，也就是说，除了访问纯文本文件之外，它们也可以用来访问二进制文件的数据。字节流类用两个类的继承来定义，在类继承树的顶层是两个抽象类：InputStream（输入流）和 OutputStream（输出流）。InputStream 定义了字节型数据输入的公共方法，OutputStream 定义了字节型数据输出的公共方法。

这两个抽象类由 Object 类扩展而来，是所有字节输入流和输出流的基类，抽象类不能直接创建流对象，而由其所派生出来的子类可以完成各种功能以及读取和写入各种设备的细节操作，提供读写不同数据的操作，图 7-2 展示了这些类之间的关系。

InputStream 类提供的方法如表 7-1～表 7-3 所示。

表 7-1 InputStream 读取数据的方法

int read( )	//读取一个字节，返回值为所读的字节
int read(byte b[ ])	//读取多个字节，放置到字节数组 b 中，通常读取的字节数量为 b 的长度，返回值为实际读取的字节的数量
int read(byte b[], int off, int len)	//读取 len 个字节，放置到以下标 off 开始字节数组 b 中，返回值为实际读取的字节的数量
int available( )	//返回值为流中尚未读取的字节的数量
long skip(long n)	//读指针跳过 *n* 个字节不读，返回值为实际跳过的字节数量

表 7-2 InputStream 标记方法

void mark(int readlimit)	//记录当前读指针所在位置，readlimit 表示读指针读出 readlimit 个字节后所标记的指针位置才失效
void reset( )	//把读指针重新指向用 mark 方法所记录的位置
boolean markSupported( )	//当前的流是否支持读指针的记录功能

表 7-3 InputStream 输出数据方法

void write(int b)	//往流中写一个字节 b
void write(byte b[ ])	//往流中写一个字节数组 b
void write(byte b[ ], int off, int len)	//把字节数组 b 中从下标 off 开始，长度为 len 的字节写入流中

关闭流：

close( )：流操作完毕后必须关闭。

flush( )：刷空输出流，并输出所有被缓存的字节（由于某些流支持缓存功能，该方法将把缓存中所有内容强制输出到流中）。

7.2.2 输入字节数据

在实际开发中，多使用字符流从控制台读取数据，这样可以使程序更直接地操作字符命令，而不必将字节流转换为字符流后再操作字符。但是，在一些简单的实用程序中，可能需要处理各种非字符的键盘操作，这时就应该使用字节流。

由于标准输入 System.in 是 InputStream 类的一个实例，因此可以直接用 System.in 操作

InputStream 类的方法。从表 7-1 中可以看到，InputStream 类中只定义了一个用于输入操作方法 read（），用它可以读取字节型数据。因此，语句 System.in .read（）的功能就是从系统标准输入流读取字节型数据。如图 7-4 所示。

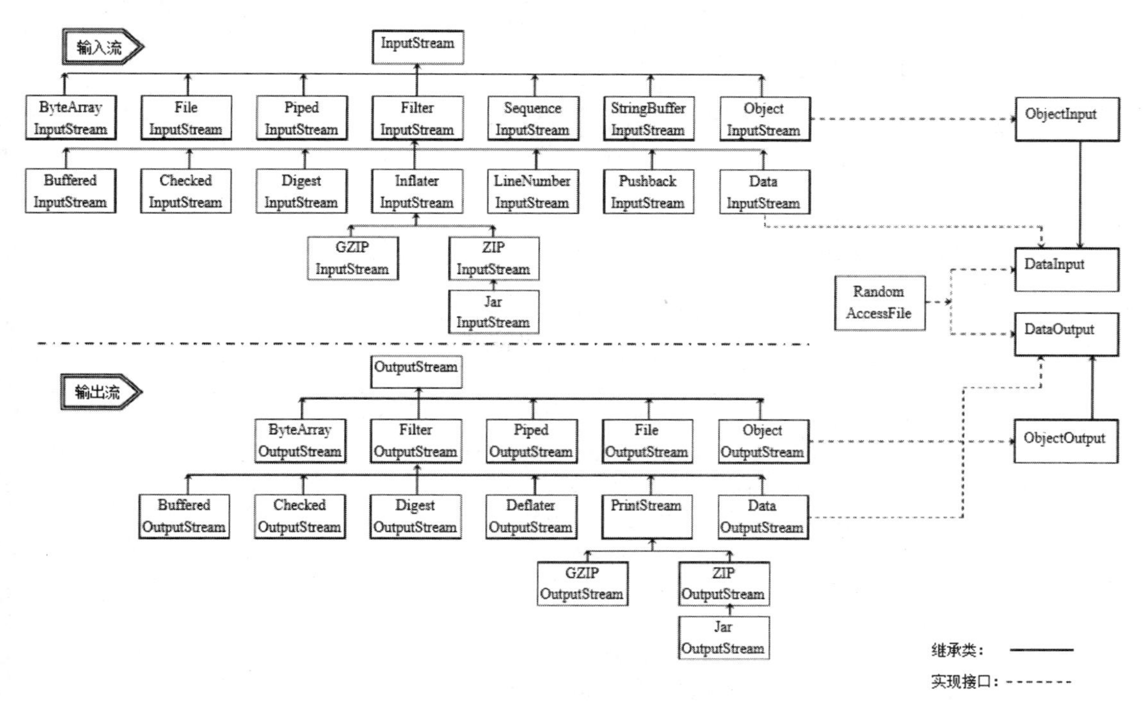

图 7-4 字节流类层次图

另外，Java 定义了字节流的子类文件输入/输出流（FileInputStream 和 FileOutputStream），它们专门用来处理磁盘文件的读写操作。它们常用的构造方法如表 7-4 和表 7-5 所示。

表 7-4 FileInputStream 的构造方法

public FileInputStream(File file)throws FileNotFoundException	根据 File 对象来创建一个 FileInputStream 类的对象
public FileInputStream(String name)throws FileNotFoundException	根据文件名称来创建一个可供读取的输入流对象

表 7-5 FileOutputStream 的构造方法

public FileOutputStream(File file)throws FileNotFoundException	根据 File 对象来创建一个 FileOutputStream 类的对象
public FileOutputStream(String name) throws FileNotFoundException	根据文件名称来创建一个可供写入数据的输出流对象，原先的文件会被覆盖

通常，FileInputStream 和 FileOutputStream 经常配合使用以实现对文件的存取，常用于二进制文件的操作。输入流 FileInputStream 中的 read（）方法按照单个字节顺序读取数据源中的数据，每调用一次，按照顺序从文件中读取一个字节，然后将该字节以整数（0～255 的一个整数）形式返回。如果到达文件末尾时，read（）返回-1。创建 Fil InputStream 对象时，若所指定的文件不存在，则会产生一个 FileNotFoundException 异常。

输入流 FileOutputStream 中的 write（）方法将字节写到输出流中。虽然 Java 在程序结束时会自动关闭所有打开的文件，但是在流操作结束后显式地关闭流仍然是一个编程的良好习惯。输入/输出流中均提供了close（）方法来显式地关闭流的操作。File OutputStream 对象的创建不依赖于文件是否存在。如果该文件存在，则它是一个目录，而不是一个常规文件；如果该文件不存在，则无法创建它；如果因为某些原因而无法打开文件，将会产生一个FileNotFoundException 异常。

✧ 编码实施

```
import java.io.*;
public class Example02 {
public static void main(String[] args) {
int a=0;
FileInputStream in = null;
FileOutputStream out = null;
try{
in = new FileInputStream("D://01readme.txt");
out = new FileOutputStream("D://02readme.txt");
while((a = in.read())!=-1){
out.write(a);
}
in.close();
out.close();
}catch(FileNotFoundException e){
System.out.println("找不到指定文件");
System.exit(-1);
}catch(IOException e){
System.out.println("文件复制错误");
System.exit(-1);
}
System.out.println("文件已复制");
try{
in = new FileInputStream("D://02readme.txt");
while((a = in.read())!=-1){
System.out.print((char)a);
}
}catch(IOException e){
System.out.println("文件打开错误");
System.exit(-1);
}
}
}
```

✧ 调试运行

第 5 行和第 6 行定义的是字节流，字节流读写文件的单元是字节，所以它不但可以读写文本文件，也可以读写图片、声音、影像文件，这个特点非常有用，因为我们可以把这种文件变成流，然后在网络上传输。

第 10～12 行从输入流中循环读取一个字符，并写入输出流，如果到达文件结尾，则返回−1，结束循环。

第 15 行、第 17 行中对可能发生的异常进行了捕获并处理：一个是创建输入流对象时，可能引发 FileNotFoundException 异常；另一个是循环读取文件中的内容时，可能引发 IOException。

第 23～31 行是将文本文件 02readme.txt 的内容在控制台中输出。

✧ 维护升级

如果读取 02readme.txt 时含有汉字，通过字节流则不能正常显示，将会出现一堆乱码。这是因为一个汉字占两个字节，而字节流读取的内容是以一个字节为单位，因此不能正确显示汉字。针对这种情况，Java 中定义了字符流。

7.2.3　字符流类

从 JDK1.1 开始，java.io 包中加入了专门用于字符流处理的类，Reader 和 Writer 是所有字符流的基类，属于抽象类，它们的子类为基于字符的输入/输出处理提供了丰富的功能。图 7-5 展示了字符流类派生的若干具体子类。

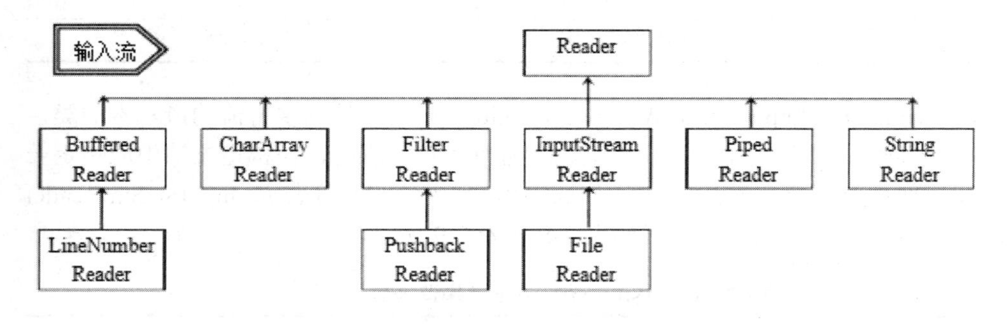

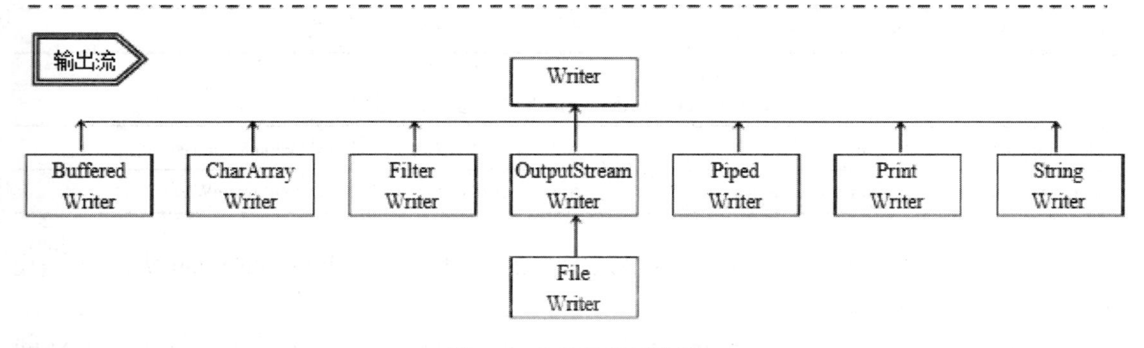

图 7-5　字符流类层次图

同类 InputStream 和 OutputStream 一样，Reader 和 Writer 也是抽象类，只提供了一系列用于字符流处理的接口。它们的方法与类 InputStream 和 OutputStream 类似，只不过其中的参数换成字符或字符数组。

字符流是以一个字符（两个字节）的长度为单位来进行数据处理，并进行适当的字符编码转换处理。

表 7-6 和表 7-7 列出了字符输入/输出流的常用方法，所有这些方法在发生错误时都会抛出 IOException 异常，Reader 和 Writer 两个抽象类定义的方法都可以被它们所有的子类继承。

表 7-6 Reader 类的常用方法

void close()	关闭输入流。如果试图继续读取，将产生一个 IOException
int read()	从输入流读取一个字符。如果到达文件结尾，则返回-1
int read(char[] cbuf)	从输入流中将指定个数的字符读入数组 cbuf 中，并返回读取成功的实际字符数目。如果到达文件结尾，则返回-1
int read(char[] cbuf, int off, int len)	从输入流中将 len 个字符从 cbuf[off]位置开始读入数组 cbuf 中，并返回读取成功的实际字符数目。如果到达文件结尾，则返回-1
boolean ready()	通知此流是否已准备好被读取

表 7-7 Writer 类的常用方法

void flush()	强制输出流中的字符输出到指定的输出流
void write(char[] cbuf)	将一个完整的字符数组写入输出流中
void write(int c)	将一个字符写入输出流中
void write(String str)	写入一个字符串到输出流中
void write(String str, int off, int len)	将指定字符串 str 中从偏移量 off 开始的 len 个字节写入输出流

InputStreamReader 和 OutputStreamWriter 是 java.io 包中用于处理字符流的最基本的类，用来在字节流和字符流之间作为中介。使用这两者进行字符处理时，在构造方法中应指定一定的平台规范，以便把以字节方式表示的流转换为特定平台上的字符表示。InputStreamReader 和 OutputStreamWriter 的构造函数的格式如表 7-8 和表 7-9 所示。

表 7-8 InputStreamReader 类构造函数

InputStreamReader(InputStream in)	创建一个使用默认字符集的输入流
InputStreamReader(InputStream in, String enc)	创建一个使用指定的字符集的输入流

表 7-9 OutputStreamWriter 类构造函数

OutputStreamWriter(OutputStream out)	创建一个使用默认的字符编码的 OutputStreamWriter
OutputStreamWriter(OutputStream out, String enc)	创建一个使用指定的字符集的 OutputStreamWriter

Java 定义了两个字符流子类的文件输入/输出流（FileReader 类和 FileWriter 类），专门用来处理磁盘文件的读写操作。

FileReader 继承自 InputStreamReader 类，而 InputStreamReader 类又继承自 Reader 类，因此 Reader 类与 InputStreamReader 类所提供的方法均可供 FileReader 所创建的对象使用。

要使用 Filereader 类读取文件，必须先调用 FileReader()构造函数 FileReader 类的对象，再利用它来调用 read（）方法。如果创建输入流时对应的磁盘文件不存在，则抛出 FileNotFoundException 异常，因此在创建时需要对其进行捕获或者继续向外抛出。

FileReader（）构造函数的格式如表 7-10 所示。

表 7-10　FileReader 类构造函数

public FileReader (File file) throws FileNotFoundException	根据 File 对象来创建一个可读取字符的输入流对象
public FileReader (String name)throws FileNotFoundException	根据文件名称来创建一个可读取字符的输入流对象

FileWrite 类继承自 OutputStreamWrite 类，而 OutputStreamWriter 类又继承自 Writer 类，因此 Writer 与 OutputStreamReader 类所提供的方法均可供 FileWriter 类所建的对象使用。要使用 FileWriter 类将数据写入文件，必须先调用 FileWriter（）构造函数创建 FileWriter 类的对象，再利用它来调用 write（）方法。FileWriter 对象的创建不依赖于文件存在与否。在创建文件之前，FileWriter 将在创建对象时打开它来作为输出。

FileWriter（）构造函数的格式如表 7-11 所示。

表 7-11　FileWriter 类构造函数

public FileWriter (File file) throws IOException	根据 File 对象来构造一个 FileWriter 对象
public FileWriter (String name) throws IOException	根据文件名称来创建一个可供写入字符数据的输出流对象，原先的文件会被覆盖

下面的程序演示了通过字符流类对文本文件的复制。

【例 7-1】 利用字符流实现文件的复制过程。

```
import java.io.*;
public class FileRead{
  public static void main(String args[])throws IOException{
    File file = new File("Hello1.txt");
    // 创建文件
    file.createNewFile();
    // creates a FileWriter Object
    FileWriter writer = new FileWriter(file);
    // 向文件写入内容
    writer.write("This\n is\n an\n example\n");
    writer.flush();
    writer.close();
    //创建 FileReader 对象
    FileReader fr = new FileReader(file);
    char [] a = new char[50];
    fr.read(a); // 从数组中读取内容
    for(char c : a)
       System.out.print(c); // 一个个打印字符
    fr.close();
  }
}
```

由于 FileReader 是以两个字节为单位读取文件中的内容，因此即使文件中有汉字，依然能够正确显示在屏幕上。

字节流在进行 IO 操作的时候，直接针对的是操作数据终端（例如，文件），而字符流操作的时候不是直接针对于终端，而是针对于缓存区（理解为内存）操作，之后再由缓存区操作终端（例如文件），属于间接操作。

7.2.4 过滤流

前面所学习的字符流和字节流提供的读取文件方法，只能一次读取一个字节或字符。如果要读取整数值、双精度或字符串，需要一个过滤流（Filter Streams）来包装输入流。使用过滤流类就可以读取整数值、双精度或字符串，而不仅仅是字节或字符。过滤流必须以某一个节点流作为流的来源，必须首先把过滤流连接到某个输入/输出流上，通常通过在构造方法的参数中指定所要连接的输入/输出流来实现。例如：

```
FilterInputStream(InputStream in);
FilterOutputStream(OutputStream out);
```

过滤流可以实现不同功能的过滤，例如缓冲流（BufferedInputStream、BufferedOutputStream 和 BufferedReader、BufferedWriter）可以利用缓冲区暂存数据，用于提高输入/输出处理的效率；数据流（DataInputStream 和 DataOutputStream）支持按照数据类型的大小来读写二进制文件。

过滤流分为面向字节和面向字符两种，例如面向字符的有 BufferedReader 类和 BufferedWriter 类，面向字节的有 DataInputStream 类和 DataOutputStream 类，其构造函数的格式如表 7-12 和表 7-13。

表 7-12 BufferedReader 类构造函数

public BufferedReader (Reader in) throws IOException	创建缓冲区字符输入流
public BufferedReader (Reader in ,int size)throws IOException	创建缓冲区字符输入流，并设置缓冲区大小
public String readLine()throws IOException	读取一行字符串

表 7-13 BufferedWriter 类构造函数

public BufferedWriter(Writer out)	创建缓冲区字符输出流
Void write（int c）thorws IOException	写入单一字符
Void write（char []cbuf,int off,int len）thorws IOException	写入字符数据的某一部分
Void write（String s,int off,int len）thorws IOException	写入字符串的某一部分
public void newLine()throws IOException	写入一个行分隔符
void flush()	刷新该流中的缓冲，将缓冲数据写到目的文件中去

【例 7-2】利用过滤流实现文件的复制过程。

```
import java.io.*;
public class TestWriter
{
    // 功能:读取E:/Test.txt文件的内容(一行一行读),并将其内容写入E:/Jack.txt中
    // 知识点:java读文件、写文件---<以字符流方式>
    public static void main(String[] args)
    {
        try
        {
            // 创建FileReader对象，用来读取字符流
            FileReader fr = new FileReader("E:/Test.txt");
            // 缓冲指定文件的输入
```

```
            BufferedReader br = new BufferedReader(fr);
            // 创建 FileWriter 对象，用来写入字符流
            FileWriter fw = new FileWriter("E:/Jack.txt");
            // 将缓冲对文件的输出
            BufferedWriter bw = new BufferedWriter(fw);
            // 定义一个 String 类型的变量,用来每次读取一行
            String myreadline;
            while (br.ready())
            {
                // 读取一行
                myreadline = br.readLine();
                // 写入文件
                bw.write(myreadline);
                bw.newLine();
                // 在屏幕上输出
                System.out.println(myreadline);
            }
            // 刷新该流的缓冲
            bw.flush();
            bw.close();
            br.close();
            fw.close();
            br.close();
            fr.close();

        } catch (IOException e)
        {
            e.printStackTrace();
        }
    }
}
```

任务 3　序列化对象

✧ 需求分析

在进行面向对象编程时，将数据与相关的操作封装在某一个类中。例如，用户的注册信息以及对用户信息的编辑、读取等操作被封装在一个类中。

1．需求描述

在实际应用中，需要将整个对象及其状态一并保存到文件中或者用于网络传输，同时又能够将该对象还原成原来的状态。这种将程序中的对象写进文件，以及从文件中将对象恢复出来的机制就是所谓的对象序列化。

2．运行结果

运行结果如图 7-6 所示。

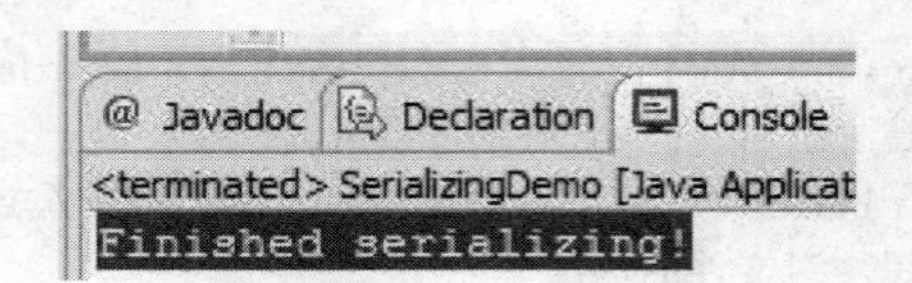

图 7-6 序列化对象

✧ 知识准备

1．技能解析

在 Java 中，对象序列化是通过 java.io.Serializable 接口和对象流类 ObjectInputStream 、ObjectOutputStream 来实现的。

2．知识解析

Java 平台允许我们在内存中创建可复用的 Java 对象，但一般情况下，只有当 JVM 处于运行时，这些对象才可能存在，即这些对象的生命周期不会比 JVM 的生命周期更长。但在现实应用中，就可能要求在 JVM 停止运行之后能够保存（持久化）指定的对象，并在将来重新读取被保存的对象。Java 对象序列化就能够帮助我们实现该功能。

7.3.1 对象序列化

使用 Java 对象序列化，在保存对象时，会把其状态保存为一组字节，在未来，再将这些字节组装成对象。必须注意的是，对象序列化保存的是对象的“状态”，即它的成员变量。由此可知，对象序列化不会关注类中的静态变量。

在 Java 中，只要一个类实现了 java.io.Serializable 接口，那么它就可以被序列化。此处将创建一个可序列化的类 Person，它实现了 Serializable 接口，它包含三个字段：name（String 类型）,age(Integer 类型),gender(Gender 类型)。另外，还重写该类的 toString()方法，以方便打印 Person 实例中的内容。

```
public class Person implements Serializable {

    private String name = null;

    private Integer age = null;

    private Gender gender = null;

    public Person() {
        System.out.println("none-arg constructor");
    }

    public Person(String name, Integer age, Gender gender) {
        System.out.println("arg constructor");
        this.name = name;
        this.age = age;
        this.gender = gender;
    }
```

```
        public String getName() {
            return name;
        }

        public void setName(String name) {
            this.name = name;
        }

        public Integer getAge() {
            return age;
        }

        public void setAge(Integer age) {
            this.age = age;
        }

        public Gender getGender() {
            return gender;
        }

        public void setGender(Gender gender) {
            this.gender = gender;
        }

        @Override
        public String toString() {
            return "[" + name + ", " + age + ", " + gender + "]";
        }
    }
```

SimpleSerial 是一个简单的序列化程序，它先将一个 Person 对象保存到文件 person.out 中，然后再从该文件中读出被存储的 Person 对象，并打印该对象。

```
    public class SimpleSerial {

        public static void main(String[] args) throws Exception {
            File file = new File("person.out");

            ObjectOutputStream    oout    =    new    ObjectOutputStream(new
FileOutputStream(file));
            Person person = new Person("John", 101, Gender.MALE);
            oout.writeObject(person);
            oout.close();

            ObjectInputStream oin = new ObjectInputStream(new FileInputStream
(file));
            Object newPerson = oin.readObject(); // 没有强制转换到 Person 类型
            oin.close();
            System.out.println(newPerson);
```

```
        }
    }
```

上述程序的输出的结果为：

```
arg constructor
[John, 31, MALE]
```

此时必须注意的是，当重新读取被保存的 Person 对象时，并没有调用 Person 的任何构造器，看起来就像是直接使用字节将 Person 对象还原出来的。

当 Person 对象被保存到 person.out 文件中之后，我们可以在其他地方去读取该文件以还原对象，但必须确保该读取程序的 CLASSPATH 中包含有 Person.class（哪怕在读取 Person 对象时并没有显示地使用 Person 类，如上例所示），否则会抛出 ClassNotFoundException。

7.3.2 Serializable 的作用

为什么一个类实现了 Serializable 接口，它就可以被序列化呢？在上节的示例中，使用 ObjectOutputStream 来持久化对象，在该类中有如下代码：

```
private void writeObject0(Object obj, boolean unshared) throws IOException {
    ...
    if (obj instanceof String) {
        writeString((String) obj, unshared);
    } else if (cl.isArray()) {
        writeArray(obj, desc, unshared);
    } else if (obj instanceof Enum) {
        writeEnum((Enum) obj, desc, unshared);
    } else if (obj instanceof Serializable) {
        writeOrdinaryObject(obj, desc, unshared);
    } else {
        if (extendedDebugInfo) {
            throw new NotSerializableException(cl.getName() + "\n"
                + debugInfoStack.toString());
        } else {
            throw new NotSerializableException(cl.getName());
        }
    }
    ...
}
```

从上述代码可知，如果被写对象的类型是 String，或数组，或 Enum，或 Serializable，那么就可以对该对象进行序列化，否则将抛出 NotSerializableException。

✧ 编码实施

1. 创建序列化对象：

```
class User implements Serializable{
    private static final long serialVersionUID = 1L;
    //transient关键字，不被序列化
    transient String username;
    String password;
    public User(String username, String password){
```

```
            this.username = username;
            this.password = password;
        }
    }
```

2．通过 ObjectOutputStream 和 ObjectInputStream 的方法实现序列化和反序列化：

```
    public static void main(String[] args)  throws IOException, ClassNotFound
Exception {
        //以下是实现对象序列化
        //创建序列化对象
        User user = new User("arthinking", "arthinking");
        //创建文件字节输出流
        FileOutputStream fos = new FileOutputStream("D:/arthinking.txt");
        //通过文件字节输出流构造对象输出流
        ObjectOutputStream oos = new ObjectOutputStream(fos);
        //写出对象
        oos.writeObject(user);
        oos.close();

        //以下是实现对象反序列化
        FileInputStream fis = new FileInputStream("D:/arthinking.txt");
        ObjectInputStream ois = new ObjectInputStream(fis);
        //不会调用任何构造方法，而是根据反序列化出来的对象状态信息构造对象
        User user2 = (User) ois.readObject();
        //username 不被序列化，这里输出为 null
        System.out.println(user2.username);
        System.out.println(user2.password);
    }
```

✧ 调试运行

1．第 1 行定义 User 类，可实现序列化。
2．第 14 行实例化 User 类。
3．第 16 行创建对象输出流的对象并指向文件“**D:/ arthinking.txt**”。
4．第 20 行将对象中的数据写入对象输出流。
5．第 21 行关闭对象输出流。
6．第 24 行保存对象的文件名。
7．第 27 行从输入流中读取对象。

✧ 维护升级

如果仅仅只是让某个类实现 Serializable 接口，而没有其他任何处理的话，则就是使用默认序列化机制。使用默认机制，在序列化对象时，不仅会序列化当前对象本身，还会对该对象引用的其他对象也进行序列化，同样地，这些其他对象引用的另外对象也将被序列化，以此类推。所以，如果一个对象包含的成员变量是容器类对象，而这些容器所含有的元素也是容器类对象，那么这个序列化的过程就会较复杂，开销也较大。

项目实训与练习

一、基本概念题

1．什么是流？什么是输入流？什么是输出流？

2．字节流与字符流有什么区别？

3. 如果准备读取一个文件的内容，应当使用FileInputStream流还是FileOutputStream流？

4．什么是java的标准输入输出流？

5．ByteArrayOutputStream流怎样获取缓冲区中的内容？

6．怎样使用RandomAccessFile流将一个文本文件倒置读出？

7．File类有哪些主要作用？它的构造方法有哪些？

8．RandomAccessFile（随机存取文件）类的主要用途是什么？它与File类有什么区别？

二、程序设计题

1．编写程序，在键盘上随机输入一串字符，在屏幕上显示出来。

2．编写程序，在键盘上输入字符，分别用字节流和字符流两种方式存入到文件中。

3．编写程序，读出一个文件，在屏幕上显示出来。

4．编写程序，用命令行参数输入一个文件名，判断其是否存在，若存在，显示其大小，创建时间，是否只读。

5．编写程序，将一个文件中的内容添加到另外一个文件的尾部。

6．编写java程序，设计一个运动员信息类Sport，要求将运动员信息类Sport的实例写入到一个文件里，然后读出并显示。

Java 的分身术：多线程机制

项目目标

本章首先介绍java线程的运行机制，然后介绍多线程的基本概念与创建、启动方法，以及如何对多个线程进行调度、同步和通信的基本知识。

项目内容

用 Java 的多线程实现时钟显示器的制作，利用线程调度机制模拟医院排队就诊程序。

任务1　时钟显示器的多线程实现

多线程机制是java语言的又一重要特征，多线程是指在操作系统中同时运行几个应用程序，每个应用程序占用一个进程在 CPU 中交替执行。使用多线程技术可以提高计算机资源的利用率，共享系统开销，提高程序执行速度。

✧ 需求分析

时钟程序运行时是多个线程交错运行的，每隔一秒调动下一个时间显示片。对于程序员来说，在设计程序时要注意给每个线程执行的时间和机会，主要是通过让线程睡眠的方法来让当前线程暂停执行，然后由其他线程来获得执行的机会。

1．需求描述

运用 Java 多线程技术编写一个电子时钟的应用程序 Clock，运行程序时会显示系统的当前日期和时间，并且每隔 1s 后会自动刷新显示当前日期和时间。

2．运行结果

运行结果如图 8-1 所示。

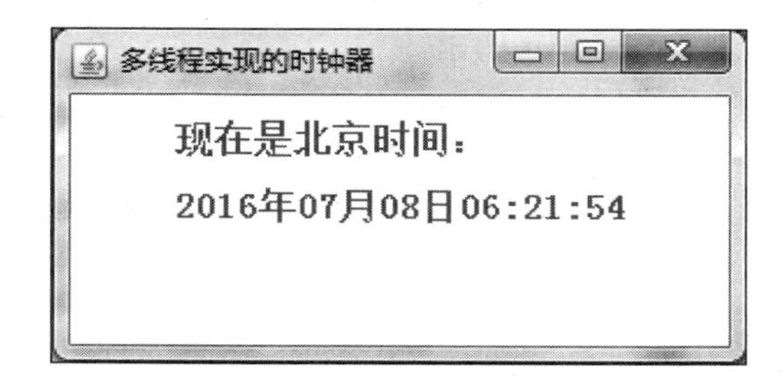

图 8-1　电子时钟的运行结果

✧ 知识准备

8.1.1 Java中的多线程机制

现在的计算机操作系统中多采用的是多任务和分时设计，进程的理念是实现多任务的一种方式。进程是指运行中的应用程序，每一个进程都有自己独立的内存空间。一个应用程序可以同时启动多个进程。线程是指进程中的一个执行流程。一个进程可以由多个线程组成，就是说一个进程中可以同时运行多个不同的线程，它们分别执行不同的任务。

以往我们开发的程序大多是单线程的，即一个程序只有一条从头至尾的执行线索。然而现实世界中的很多过程都具有多条线索同时动作的特性。例如，我们可以一边看电视一边扫地，再如一个网络服务器可能需要同时处理多个客户机的请求等。多线程的应用范围很广，通常可以将一个程序任务转换成多个独立并行运行的子任务。WEB浏览器就是一个多线程应用程序，当下载一个应用程序或图片时，可以同时播放动画或声音，或者同时在后台打印或进行其他工作。

尤其在网络编程中，多线程编程是非常有用的，例如网络传输较慢时，用户输入速度慢时，这时可以分别用两个独立的线程来完成这些功能，却不影响正常的显示和其他功能。

多线程是指在单个程序中可以同时运行多个不同的线程，执行不同的任务，它是实现并发机制的一种有效手段。从逻辑的观点来看，多线程意味着一个程序的多行语句同时执行，可以简化应用程序设计，增进程序的交互性，提供更好的GUI和服务器功能。

本任务是创建一个Java多线程应用程序，使用Canvas类来创建一个画板,使用Graphics类在画板中绘制出系统当前日期和当前时间，使用Thread类创建一个线程。并重写线程中的run方法来实现每隔一秒钟刷新显示系统的当前日期和当前时间。

8.1.2 线程与进程

线程是程序中一个单一的顺序控制流程，它是程序运行的基本执行单元。线程是比进程还小的单位。线程有它自己的入口和出口，以及一个顺序执行的序列。线程不能独立存在，必须存在于进程中。 线程最大的一个特性是并发执行，多个并发执行的线程称为多线程。

使用线程将会充分利用CPU资源，简化编程模型和异步事件的处理，使GUI更有效率，从而达到节约成本的方式来改善应用程序。

线程与进程的主要区别在于：线程的划分尺度小于进程； 进程在执行过程中拥有独立的内存单元，而多个线程只能共享进程的内存单元；在执行过程中，每个独立的线程有一个程序运行的入口、顺序执行序列和程序的出口。但是线程不能够独立执行，必须依存于应用程序中，由应用程序提供多个线程执行控制；从逻辑角度来看，多线程的意义在于一个应用程序中，有多个执行部分可以同时执行。但操作系统并没有将多个线程看做多个独立的应用，来实现进程的调度和管理以及资源分配；一个线程可以创建和撤销另一个线程；同一个进程的多个线程之间可以并发执行。

8.1.3 线程生命周期

同进程一样，一个线程也有从创建、启动到消亡的过程，整个过程称为一个生命周期。

每个线程都和生命周期相关联，在此期间的任何时刻，总是处于创建、就绪（可运行）、运行、阻塞、死亡这 5 种状态中的某一种状态。 并且生命周期中的这几种状态可以互相转化，通过线程的控制与调度可使线程在这几种状态间转化，转化状态如图 8-2 所示。

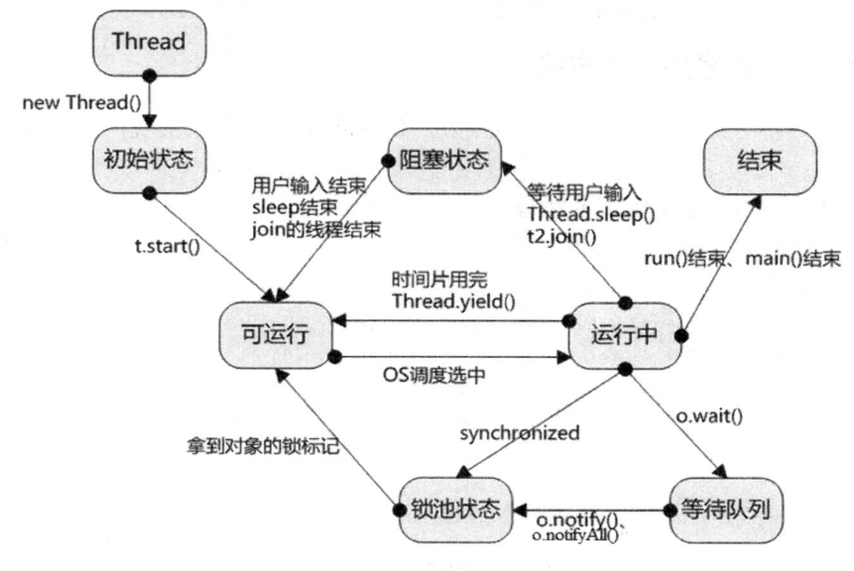

图 8-2　Java 线程状态转换

（1）创建状态：用 new 运算符创建一个 Thread 类或子类的实例对象时，形成的新线程就进入创建状态，但此时还未给这个线程分配任何资源，没有真正执行它。

（2）就绪状态： 处于就绪状态的线程已经具备了运行条件，但还没有分配到 CPU，因而将进入线程队列，等待系统为它分配 CPU。此时，start()方法分配运行这个线程所需的系统资源，安排其运行，并调用线程体的 run()方法，start()方法返回后，线程处于可运行状态也称为可运行状态。

（3）运行状态：当就绪状态的线程被调度并获得 CPU 处理器资源时，便进入运行状态，这时便开始顺序执行自己的 run 方法中的代码，直到调用其他方法而终止，或等待某资源而阻塞，或完成任务而死亡。运行状态表示线程拥有了对处理器的控制权，其代码正在运行，除非运行过程的控制权被另一优先级更高的线程抢占，否则这个线程将一直持续到运行完毕。

（4）阻塞状态 ：由于某种原因使得运行中的线程不能继续执行，该线程进入阻塞态。处于运行状态的线程在某些情况下，如执行了 sleep（睡眠）方法，或等待 I/O 设备等资源，将让出 CPU 并暂时终止自己的运行，进入阻塞状态。在阻塞状态的线程不能进入就绪队列，只有当引起阻塞的原因消除时，如睡眠时间已到或等待的 I/O 设备空闲下来，线程转入就绪状态。

此时线程不会被分配 CPU 时间，无法执行。Java 中提供了大量的方法来阻塞线程。

发生以下情况之一，线程进入阻塞状态：

- 调用了该线程的 sleep()休眠方法。
- 调用了 wait()等待方法，使得线程进行阻塞状态。它有两种格式：一种是允许指定以毫秒为单位的一段时间内作为参数，该格式可以用 notify()方法被调用或超出指定时间时，线程可重新进入可运行状态；另一种格式没有格式，该格式必须是 notify()

方法被调用后才能使线程进行可运行状态。

➢ 调用了 suspend()挂起方法。

➢ 该线程正在等待 I/O 操作完成。

（5）终止状态：死亡状态是线程生命周期中的最后一个阶段，表示线程已退出运行状态。

8.1.4 多线程的实现方式

Java 中有两种方式实现多线程：一是通过 Thread 类的子类来实现，二是通过 Runnable 接口来实现。

1．通过继承 Thread 类实现多线程

【例 8-1】Thread 类实现的多线程编程。

```
public class Threaddemo_1 {
public static void main(String[] args) throws InterruptedException {
HelloThread ht = new HelloThread("Hello ");
ht.start();              //启动线程
Thread.sleep((int)Math.random()*100);   //主线程休眠0~100ms的随机数
System.out.println("Java! ");
}
}
class HelloThread extends Thread{
public HelloThread(String s){        //构造方法
super(s);              //调用父类的构造方法，线程名为s
}  //必须覆盖Thread类的run()方法，线程启动后将执行该方法
public void run(){
System.out.print(getName());        //打印线程名
}
}
```

一个 Thread 类的一个实例对象就是 Java 程序的一个线程，所以 Thread 类的子类的实例对象也是 Java 程序的一个线程。因此构造 Java 程序可以通过构造类 Thread 的子类的实例对象来实现。构造类 Thread 的子类主要目的是为了让线程类的实例对象能够完成线程程序所需要的功能。

通过这种方法构造出来的线程在程序执行时的代码被封装在类 Thread 或其子类的成员方法 run 中。为了使新构造出来的线程能完成所需要的功能。新构造出来的线程类应覆盖 Thread 类的成员方法 run()。

线程的启动或运行并不是调用成员方法 run()。而是调用成员方法 start()达到间接调 run()方法，线程的运行实际上就是执行线程的成员方法 run()。

Thread 类提供的实现多线程编程的基本控制方法有：

（1）run（）方法。线程的所有活动都是通过线程体 run()方法来实现的。在一个线程被建立并初始化以后，Java 运行时系统就自动调用 run()方法，所以实现线程的核心是实现 run()方法，它也是线程开始执行的起始点。

（2）start()方法 。线程调用该方法启动一个线程，使之从新建状态进入就绪队列排队，一旦轮到它来使用 CPU 资源时，就可以脱离创建它的主线程独立开始自己的生命周期。

（3）sleep（）方法。使线程睡眠一段时间，单位为毫秒。线程的调度执行是按照其优先级的高低顺序进行的，当高级线程不完成，即未死亡时，低级线程没有机会获得处理器。有时，优先级高的线程需要优先级低的线程做一些工作来配合它，或者优先级高的线程需要完成一些费时的操作，此时优先级高的线程应该让出处理器，使优先级低的线程有机会执行。为达到这个目的，优先级高的线程可以在它的 run()方法中调用 sleep()方法来使自己放弃处理器资源，休眠一段时间。休眠时间的长短由 sleep()方法的参数决定。如果线程在休眠时被打断，JVM 就抛出 InterruptedException 异常。因此，必须在 try-catch 语句块中调用 sleep()方法。

（4）isAlive()方法。public final boolean isAlive()测试线程是否处于活动状态。如果线程已经启动且尚未终止，则为活动状态。 返回：如果该线程处于活动状态，则返回 true；否则返回 false。

（5）interrupt()方法。intertupt()方法经常用来“吵醒”休眠的线程。当一些线程调用 sleep()方法处于休眠状态时，一个使用 CPU 资源的其他线程在执行过程中，可以让休眠的线程分别调用 interrupt()方法“吵醒”自己，即导致休眠的线程发生 InterruptedException 异常，从而结束休眠，重新排队等待 CPU 资源。

（6）yield()方法。把线程移到队列的尾部。

（7）stop()方法：结束线程生命周期并执行清理工作。

（8）destroy()方法：结束线程生命周期但不做清理工作。

【例 8-2】通过继承 Thread 类创建线程，在主控程序中同时运行多个线程。

```
public class TestThread03 extends Thread
{ private static int threadCount = 0;
  private int threadNum = ++threadCount;
  private int i = 5;
  public void run()  {
    while(true)   {
     try{ Thread.sleep(5);}
     catch(InterruptedException e){ System.out.println("Interrupted"); }
     System.out.println("Thread " + threadNum + " = " + i);
     if(--i==0) return;}
   }
   public static void main(String[] args)
   { for(int i=0; i<5; i++)
     {  new TestThread03().start();  }
   }
}
```

程序运行的结果如图8-3所示。

说明：（1）run()方法必须进行覆盖，把要在多个线程中并行处理的代码放到这个函数中。

（2）在 main()方法中创建两个线程，线程被创建后不会自动执行线程，而需要调用 start()方法来启动这两个线程。

（3）main()方法本身也是一个线程，在 main()方法中产生并启动 5 个线程后，输出活动线程信息。

2. 用 Runnable 接口创建线程

由于 Java 规定类只能继承一个类，对于一个已经有父类的子类在实现线程时就不能用 Thread 类来实现了，则应采用 Runnable 接口来实现。

通过实现 Runnable 接口来编写多线程程序，只需要重写 run()方法，而且实现 Runnable 接口的类还可以继承其他类，在这一点上它与定义 Thread 类的子类实现多线程不同，也是它的优势所在。

具体实现步骤如下：

（1）定义一个实现 Runnabl 接口的类；

（2）定义方法 run().Runnable 接口中有一个空的方法 run()，实现它的类必须覆盖此方法；

（3）创建该类的一个线程对象，并以该对象为参数，传递给 Thread 类的构造方法，从而生成 Thread 类的一个对象；

（4）通过 start()方法启动线程。

```
Console
<terminated> TestThread03
Thread 4 = 3
Thread 2 = 2
Thread 1 = 2
Thread 3 = 2
Thread 5 = 2
Thread 4 = 2
Thread 2 = 1
Thread 3 = 1
Thread 1 = 1
Thread 4 = 1
Thread 5 = 1
```

图 8-3 继承 Thread 类创建线程

【例 8-3】用 Runnable 接口创建线程。

```
public class RunnableDemo_1 {
 public static void main(String[] args) throws InterruptedException{
 RunThread rt1 = new RunThread(); //定义 RunThread 对象 rt1
 RunThread rt2 = new RunThread(); //定义 RunThread 对象 rt2
  //以实现 Runnable 接口的类对象为参数，定义 Thread 对象
 Thread t1 = new Thread(rt1,"线程 1");
  Thread t2 = new Thread(rt2,"线程 2");
  System.out.println("main 开始! ");
t1.start();//启动线程 1
t2.start();//启动线程 2
  System.out.println("main 结束! ");
 }
}
class RunThread implements Runnable{
 //必须覆盖 Runnable 接口的 run()方法,线程启动后将执行该方法
@Override
 public void run() {   //打印线程运行
  System.out.println(Thread.currentThread().getName() + "运行");
                                                        //打印线程结束
  System.out.println(Thread.currentThread().getName() + "结束");
 }
}
```

例如，通过实现 Runnable 接口来创建线程。

```
public class RunnableTest{
public static void main(String args[]) {
Target first,second;
first=new Target("第一个 Runnable 线程");
second=new Target("第二个 Runnable 线程");
```

```
        Thread one,two;
        one=new Thread(first);
        two=new Thread(second);
        one.start();
        two.start();}
            }
        class Target  implements Runnable{
            String s;
        public Target(String s){
            this.s=s;
        System.out.println(s+"现在已经建立");}
        public void run(){
            System.out.println(s+"今天已经运行");
            try{Thread.sleep(1000);
            }catch(InterruptedException e){}
            System.out.println(s+"目前已经结束");}
        }
```

程序运行的结果如图8-4所示。

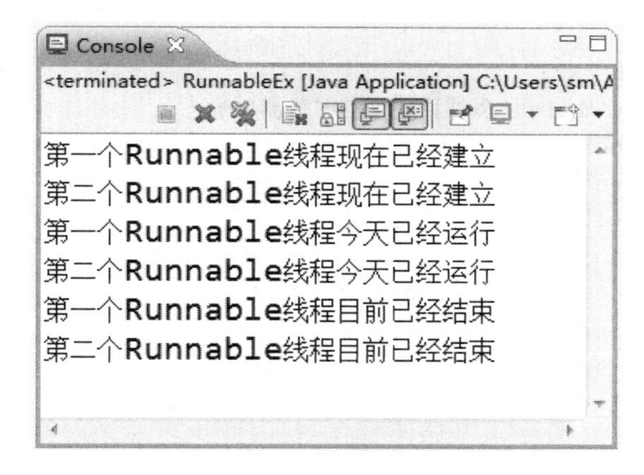

图 8-4　实现 Runnable 接口创建线程

说明：

（1）Runnable 接口只有一个方法 run()，使用 Runnable 接口创建线程子类，必须重写 Runnable 接口中的 run()方法。

（2）RunnableTest 类的对象 first 虽然有 run()方法的线程体，但其中没有 start()方法，所以 first 不是一个线程对象，而只能作为一个带有线程体的目标对象。

（3）如果线程类是继承 Thread 类，那么只需对其使用 new 来创建实例。如果是实现 Runnable 接口，就需要将类的实例传递给 Thread 构造器，即在 main()方法中，必须将对象 first、second 作为目标对象，构造出 Thread 类的线程类 first 和 second，通过调用 one.start()和 two.start()来启动线程。

✧ 编码实施

时钟程序代码如下：

```
import java.awt.BorderLayout;
import java.awt.Canvas;
import java.awt.Color;
import java.awt.Font;
import java.awt.Graphics;
import java.sql.Date;
import java.text.SimpleDateFormat;
import java.util.Calendar;
import javax.swing.JFrame;
import javax.swing.JPanel;
class Clock extends Canvas {
    JFrame frame=new JFrame();
    JPanel conPane;
    String time;
    int i=0;
    Date timer;
    public static void main(String args[]){
        Thread ClockThread= new ClockThread();
        ClockThread.start();
}
    public Clock(){
     frame.setTitle("多线程实现的时钟器");
        conPane=(JPanel)frame.getContentPane();
        conPane.setLayout(new BorderLayout());
        conPane.setSize(280,40);
        conPane.setBackground(Color.white);
        conPane.add(this,BorderLayout.CENTER);
        frame.setVisible(true);
        frame.setSize(300, 150);
        frame.setDefaultCloseOperation(JFrame.EXIT_ON_CLOSE);
    }
    public void paint(Graphics g){
        Font f=new Font("宋体",Font.BOLD,16);
        SimpleDateFormat SDF=new SimpleDateFormat("yyyy'年'MM'
       月'dd'日'HH:mm:ss");//格式化时间显示类型
        Calendar now=Calendar.getInstance();
        time=SDF.format(now.getTime());//得到当前日期和时间
        g.setFont(f);
        g.setColor(Color.red);
        g.drawString("现在是北京时间: ",45,25);
        g.drawString(time,45,55);
    }

}
class ClockThread extends Thread{
 Clock t=new Clock();
public void run(){
```

```
        while(true){
          try{
             Thread.sleep(1000);  //休眠 1 秒钟
                              }
    catch(InterruptedException e){
            System.out.println("异常");
                          }
                      t.repaint(100);
            }
     }
    }
```

✧ 调试运行

1．在项目中创建类 Clock，得到类的框架。

```
public class  Clock {
}
```

2．在 public class Clock 下面一行创建对象和输入类的属性描述：

```
JFrame frame=new JFrame();
   JPanel conPane;
Date timer;
   String time;
   int i=0;
```

3. 在 Clock 类中的 paint()方法中使用 Graphics 类的 drawString 方法显示系统的当前日期和时间。

4．定义 ClockThread 线程类，在 ClockThread 线程类中的 run 方法中循环调用 repaint() 方法来刷新显示系统的当前日期和时间。

✧ 维护升级

在电子时钟中可以加入星期的显示功能，上述代码部分修改如下：

```
public void paint(Graphics g){
      Font f=new Font("宋体",Font.BOLD,16);
      SimpleDateFormat SDF=new SimpleDateFormat("yyyy'年'MM'
      月'dd'日'HH:mm:ss");//格式化时间显示类型
      Calendar now=Calendar.getInstance();
      time=SDF.format(now.getTime());     //得到当前日期和时间
      g.setFont(f);
      g.setColor(Color.orange);
      g.drawString(time,45,25);
     //以下为计算星期的代码
 dayOfWeek = now.get(Calendar.DAY_OF_WEEK);
      switch (dayOfWeek) {
      case 1:
        g.drawString("星期日",120,60);
        break;
      case 2:
```

```
        g.drawString("星期一",120,60);
        break;
      case 3:
        g.drawString("星期二",120,60);
        break;
      case 4:
        g.drawString("星期三",120,60);
        break;
      case 5:
        g.drawString("星期四",120,60);
        break;
      case 6:
        g.drawString("星期五",120,60);
        break;
      case 7:
        g.drawString("星期六",120,60);
        break;
      }
```

修改代码后，重新运行程序，程序运行的结果如图8-5所示。

图8-5 加入星期后的电子时钟的运行结果

Java程序中的线程并发运行将共同争抢CPU资源，哪个线程抢夺到CPU资源后就开始运行，这样就会造成线程之间的冲突，为了规避这种现象，引出了线程调度线程等待的概念。

任务2 线程调度

在CPU上以某种次序执行多个线程被称为调度。线程调度的意义在于JVM应对运行的多个线程进行系统级的协调，以避免多个线程争用有限资源而导致应用系统死机或者崩溃。

✧ 需求分析

模拟医院挂号看病的流程，要求可以从两个不同候诊区进入诊区看病，且每次只能进一人，两个通道同时往诊区里进入看病，且每个大夫同时只能给一人看病。4个线程同步执行、相互协调。看病时，必须保证前面没有患者，且不能同时进入。

1．需求描述

本任务是创建一个Java多线程状态设置与线程调度应用程序，首先创建一个普通类jiuZhen，在此类中创建两个方法。第一个方法为put()方法，实现将患者放进诊区。第二个方法为get()方法，实现将有一个患者进入诊区看病。然后再创建两个线程来分别调用put()方法和 get()方法来完成进入诊区和看病的操作。

2．运行结果

运行结果如图 8-6 所示。

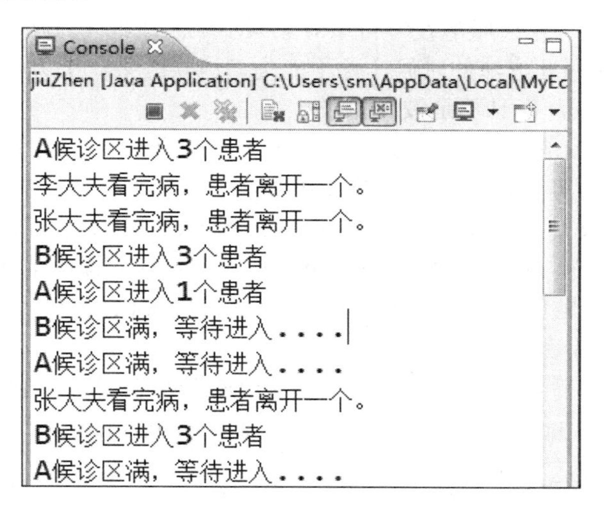

图 8-6 “挂号”的线程调度的运行结果

✧ 知识准备

8.2.1 线程的优先级

Java 虚拟机允许一个应用程序可以拥有多个同时执行的线程，哪一个线程先执行，哪一个线程后执行，取决于线程的优先级，优先级高的先执行，优先级低的后执行，优先级相同的，则遵循队列的“先进先出”原则。这就是 Java 定义的线程的优先级策略。

线程的优先级代表该线程的重要程度或紧急程度。当有多个线程同时处于可执行状态，并等待获得 CPU 时间时，Java 虚拟机会根据线程的优先级来调用线程。在同等情况下，优先级高的线程会先获得 CPU 时间，优先级较低的线程只有等排在它前面的高优先级线程执行完毕之后才能获得 CPU 资源，对于优先级相同的线程，则遵循队列的“先进先出”原则，即先进入就绪态的线程被优先分配使用 CPU 资源。

Java 线程的优先级从低到高以整数 1～10 表示，共分为 10 级，可以调用 Thread 类的 getPriority()方法获取线程的优先级和 setPriority(int newPriority)方法来改变线程的优先级。

Thread 类优先级有关的成员变量如下。

MAX_ PRIORITY：一个线程可能有的最大优先级，值为 10。

MIN_PRIORITY：一个线程可能有的最小优先级，值为 1。

NORM_PRIORITY：一个线程默认的优先级，值为 5，即 Thread. NORM_PRIORITY。

8.2.2 线程调度方法

Java 系统中的多线程会共争资源，但它们之间存在一定的关系，这就需要调度，可以通过如下三个方法实现多线程间的调度和通信。

➢ wait()：等待方法。可让当前线程放弃监视器并进入睡眠状态，直到其他线程进入同一监视器并调用 notify 为止。

wait()方法的定义是：public final void wait() throws InterruptedException

wait 的作用：释放已持有的锁,进入等待队列。

➢ notify()：唤醒方法。唤醒在等待该对象同步锁的线程（只唤醒一个，如果有多个在等待），注意的是在调用此方法的时候，并不能确切地唤醒某一个等待状态的线程，而是由JVM 确定唤醒哪个线程，而且不是按优先级。

调用任意对象的 notify()方法则导致因调用该对象的 wait()方法而阻塞的线程中随机选择的一个解除阻塞（但要等到获得锁后才真正可执行）。

notify()方法的定义是：public final native void notify()

notify 的作用：唤醒 wait 队列中的第一个线程并把它移入锁申请队列。

➢ notifyAll()：唤醒所有等待的线程，注意唤醒的是 notify 之前 wait 的线程，对于 notify 之后的 wait 线程是没有效果的。

通常，多线程之间需要协调工作：如果条件不满足，则等待；当条件满足时，等待该条件的线程将被唤醒。在 Java 中，这个机制的实现依赖于 wait/notify。

8.2.3 线程的同步

由于一个进程的线程共享同一片存储空间，当两个线程都需要同时访问同一个数据对象时，会带来严重的访问冲突问题，Java 语言提供了专门机制以解决这种冲突，确保任何时刻只能有一个线程对同一个数据对象进行操作。这套机制就是 synchronized 关键字，它有两种用法：synchronized 方法和 synchronized 块。

1．synchronized 方法

通过在方法声明中加入 synchronized 关键字来声明 synchronized 方法。例如：synchronized void put()

synchronized 方法控制对类成员变量的访问，每个类实例对应一把锁。每个 synchronized 方法都必须获得调用该方法的类实例的锁才能执行，否则所属线程阻塞。方法一旦执行，就独占该锁，直到从该方法返回时才将锁释放，然后只有被阻塞的线程才能获得该锁，重新进入可执行状态。这种机制保证了同一时刻同一个数据对象只能被一个线程操作。

2．synchronized 块

通过 synchronized 关键字来声明 synchronized 块的语法如下：

```
Synchronized(syncObject){
    //被访问控制的代码
}
```

synchronized 块中的代码必须获得 syncObject 的锁才能执行。由于可以针对任意代码块，且可任意指定上锁对象，故灵活性较高。

【例 8-4】带锁定的售票线程。

```
class saleTest extends Thread {
  private int tickets=100;
  String str=new String("");
  public void run(){
  while(true) {
    synchronized(str) {      //锁定对象
```

```
            if (tickets>0) {
                try
                  {Thread.sleep(10);
                  }
                  catch (Exception e)
                    {System.out.println(e.getMessage());}
               }
            System.out.println(Thread.currentThread().getName()+"
正在售 "+tickets--+" 票");
            }
         }
        }
      }
    }
```

✧ 编码实施

```
public class jiuZhen {
     int f=5;  //诊区里最多可以进去 5 个患者
     int n=0;  //诊区里剩余的患者数量
     int num;  //最多可以进去的患者数量
     int s=0; //已经进入的患者数量
     int z=0; //已经看完病的患者数量
     public jiuZhen(int num) {
     this.num=num;
}
synchronized void get(){
     while(n<1)  {
     try
          { System.out.println(Thread.currentThread().getName()+"空闲中……");
           wait();
          }catch (InterruptedException e){}
       }
       z=z+1;
       if (z==num) Thread.currentThread().stop();
       n=n-1;
       System.out.println(Thread.currentThread().getName()+"看完病,患者离开一个。");
       notify();
       }

     synchronized void put(){
          while(n>=f) {
           try
               { System.out.println(Thread.currentThread().getName()+"候诊
               区满，等待进入....");
                 wait();
                }catch (InterruptedException e){}
            }
     int x=(int)(Math.random()*3)+1;
```

```
            s=s+x;
            if (s>=num) Thread.currentThread().stop();
            n=n+x;
            System.out.println(Thread.currentThread().getName()+"候诊区
            进入"+x+"个患者");
         notify();
         }
      public static void main(String args[]){
         jiuZhen work=new jiuZhen(20);
       producter p1=new  producter("A",work);
       producter p2=new  producter("B",work);
       doctor c1=new  doctor("张大夫",work);
       doctor c2=new  doctor("李大夫",work);
       p1.start();
       p2.start();
       c1.start();
       c2.start();
      }
   }
   class producter extends Thread{     //创建患者线程
      jiuZhen work;
      producter(String name,jiuZhen work) {
       super(name);
       this.work=work;
      }
      public void run(){
     while (true)  {
      work.put();
        try
           { Thread.sleep(100);
           }catch (InterruptedException e){}
        }
      }
   }
     class doctor extends Thread{      //创建看病的线程
         jiuZhen work;
         doctor(String name,jiuZhen work)  {
          super(name);
         this.work=work;
         }
         public void run(){
            while (true) {
             work.get();
             try
            { Thread.sleep(1000);
            }catch (InterruptedException e){}
          }
       }
   }
```

✧ 调试运行

1．在 study 项目中创建类 jiuZhen 得到类的框架。

```
public class jiuZhen{
}
```

2．在 public class jiuZhen 下面一行输入类的属性描述：

```
int f=5;     //诊区里最多可以进去 5 个患者
int n=0;     //诊区里剩余的患者数量
int num;     //最多可以进去的患者数量
int s=0;     //已经进入的患者数量
int z=0;     //已经看完病的患者数量
```

3．在 EatApple 类中输入两个方法的定义：

```
public jiuZhen(int num) {
    this.num=num;
        }
synchronized void put(){
        ……          //参见详细实现代码
    }
    synchronized void get(){
        ……          //参见详细实现代码
    }
    public static void main(String args[]){
        ……          //参见详细实现代码
    }
```

4．定义 producter 线程类，在 producter 线程类中的 run 方法中循环调用 jiuZhen 类中 put()方法来完成放苹果的操作。

5．定义 constumer 线程类，在 constumer 线程类中的 run 方法中循环调用 jiuZhen 类中 gett()方法来完成取苹果的操作。

✧ 维护升级

可以将一次只能进入一名患者改为一次可以随机进入多个患者，上述代码部分修改如下。

```
import java.math.*;
synchronized void put(){
    while(n>=f) {
     try
        { System.out.println(Thread.currentThread().getName()+"
        等待进入诊区");
          wait();
         }catch (InterruptedException e){}
      }
        int x=(int)(Math.random()*3)+1;
      s=s+x;
      if (s>=num) Thread.currentThread().stop();
      n=n+x;
      System.out.println(Thread.currentThread().getName()+"
      诊区可以进入"+x+"个患者");
```

```
        notify();
    }
```

修改代码后，重新运行程序，程序运行的结果如图 8-7 所示。

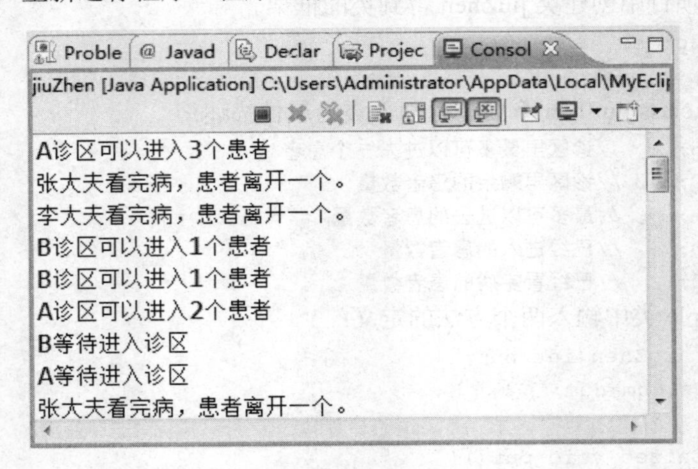

图 8-7 可以随机进入候诊区的运行结果

项目实训与练习

一、简述题

1. 简述线程和进程的关系。
2. 描述线程的生命周期。
3. 引起线程中断的原因是什么？
4. 建立线程有几种方法？
5. 怎样设置线程的优先级？
6. 线程有几种状态？
7. 简述线程如何进行调度。
8. 简述多线程之间怎样进行调度。

二、编程题

1. 试编写两个线程：其一用来计算 2~7000 间的质数及其个数；其二用来计算 2001～7000 间质数及其个数。

2. 编写一个 Applet 程序，实现一个字符串或图形的不停移动。

3. 编写程序，在主线程中创建三个线程："运货司机"、"装卸工"和"仓库管理员"。要求："运货司机"占有 CPU 资源后立刻联合线程"装卸工"，也就是让"运货司机"一直等到"装卸工"完成工作才能开车；而"装卸工"占有 CPU 资源后立刻联合线程"仓库管理员"，也就是让"装卸工"一直等到"仓库管理员"打开仓库才能开始搬运货物。

项目 9

网络编程

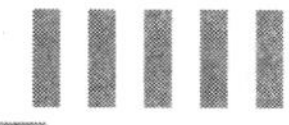

项目目标

本章的主要内容是了解网络基础知识，熟悉网络编程中的基本概念；熟悉 Java 网络开发中类的使用方法；初步理解并比较 TCP 协议与 UDP 协议两种网络编程的实现方式；能基于 TCP 协议来编写一个面向连接的网络通信程序；能分别使用 Socket 类与 ServerSocket 类来创建客户端程序与服务端程序，并实现客户端程序与服务端程序的信息交换。

项目内容

使用 java 网络技术实现一个简单的聊天室程序，可以在服务器端和客户端进行通信。

任务 1　基于 TCP 实现简单聊天室程序

✧ 需求分析

基于 TCP 的网络数据传输是一种可靠的有连接的网络数据传输。在基于 TCP 的网络程序中，运用网络通信基础知识、网络地址和套接字等技术编写一个面向连接网络应用程序，在服务器端程序会接收客户端程序发送的信息并作处理。在客户端的程序也会接收到服务器端发送的信息并作处理。

1．需求描述

本任务是基于 TCP 有连接的网络应用程序，首先要创建服务端的程序 TcpServer。在服务端的程序 TcpServer 中要创建 ServerSocket 类的实例对象，注册在服务器端进行连接的端口号。并调用 ServerSocket 类的实例对象的 accept()方法来等待客户的连接。其次是创建客户端的程序 TcpClient。在客户端的程序 TcpClient 中要创建 Socket 类的实例对象，通过该实例对象与服务器端程序建立连接，实现信息的传递。

2．运行结果

首先运行服务器端程序 TcpServer，接着运行客户端程序 TcpClient，程序运行的结果如图 9-1 所示。

✧ 知识准备

服务器首先启动，创建套接字后等待客户的连接；客户启动以后，创建套接字，然后和服务器建立连接；连接建立后，客户机和服务器可以通过建立的套接字连接进行通信。服务器和客户端可以是一台电脑的两个进程，也可以分别部署在两台电脑上。

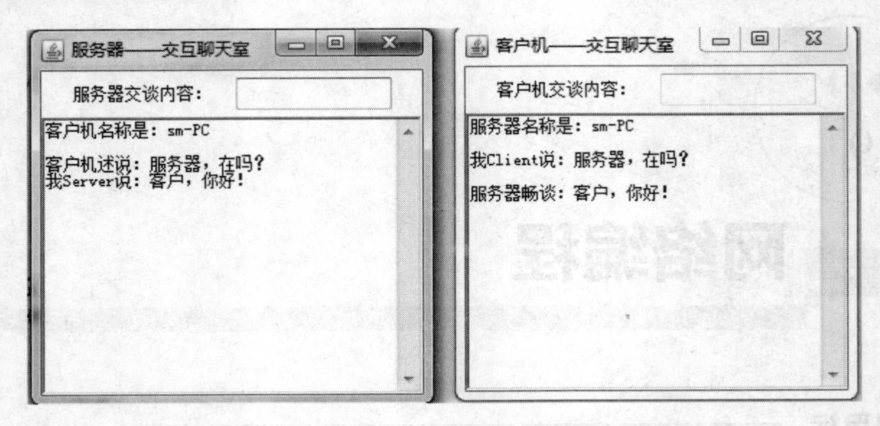

图9-1 基于TCP的一对一的Socket通信的运行结果

9.1.1 网络通信概述

1．TCP/IP协议

TCP/IP是一种计算机间的通信规则，是网络中的基本通信协议。它规定了计算机之间通信的所有细节，规定了每台计算机信息表示的格式和含义，规定了计算机之间通信所使用的控制信息，以及在接收到控制信息后应该做出的反应。TCP/IP协议是Internet中计算机之间通信时必须共同遵守的一种通信协议，是一个协议集（简称为TCP/IP协议）。

2．IP地址与域名

通信时，为了确保网络上每一台主机能够互相识别，必须给每台主机一个唯一的地址，即IP地址，用来标识主机在网上的位置。IP地址由32位二进制数构成，分为四段，每段8位，可用小于256的十进制数来表示，段间用圆点隔开。例如：192.168.10.1。

虽然采用IP地址标识Internet十分有效，但由于这种标识不方便记忆，所以引进了便于记忆的、富有含义的字符型IP地址：域名。它是统一资源定位器URL的基础，采用一种分步型层次式的命名机制。域名由若干子域构成，子域间以圆点相隔，最右边的子域是顶级域名，至右向左层次逐级降低，最左边的子域是主机名。域名的一般形式为："主机名.网络名.机构名.顶级域名"。

3．端口（port）

计算机与网络一般只有一个物理连接，网络数据通过这个连接流向计算机，但是计算机上一般有许多个进程在同时运行，为确定网络数据流向指定的进程，TCP/IP引入了端口概念。端口和IP地址为网络通信的应用程序提供了一种确定的地址标识，IP地址表示了发送端的目的计算机，而端口表明了将数据发送给目的计算机上的哪一个应用程序。

端口同一台网络计算机的一个特定进程关联，与进程建立的套接接口绑定在一起。客户程序必须事先知道自己要求的那个服务进程的IP和端口号（端口号与IP地址的组合得出一个网络套接字），才能与那个服务建立连接并进行通信，程序员在创建自己的应用程序时一般自己指定一个端口号，客户通过这个端口号连接该服务进程。客户端应用进程与服务进程一样，也有自已的的端号，通过该端口来完成客户端应用进程与服务进程进行通信。

0～1023之间的端口是用于一些知名的网络服务和应用。如80是HTTP的缺省端口号，21是文件传送服务FTP的缺省端口号。用户的普通网络应用服务应使用1024以上的端口，从而避免端口号已被另一个应用或系统服务所用。

9.1.2 URL编程

类URL代表一个统一资源定位符，它是指向互联网“资源”的指针。资源可以是简单的文件或目录，也可以是更为复杂的对象的引用，例如对数据库或搜索引擎的查询。

简单地可以把URL理解为包含：协议、主机名、端口、路径、查询字符串和参数等对象。每一段可以独立设置。

应用程序也可以指定一个“相对URL”，它只包含到达相对于另一个URL的资源的足够信息。HTML页面中经常使用相对URL。

Java语言访问网络资源是通过URL类来实现的。要使用URL进行通信，就要使用URL类创建其对象，通过调用URL类的方法完成网络通信。

1．创建URL类的对象

URL类提供用于创建URL对象构造方法有如下4个。

（1）public URL(String url)：它是使用URL的字符串来创建URL对象。

如：URL myurl=new URL（“http://www.sit.edu.cn/”）；

（2）public URL（URL baseURL, String relativeURL）。

baseURL是绝对路径，relativeURL是相对位置。如：

URL myWeb=new URL（“http://www.dhu.edu.cn/”）；

URL myMat=new URL（myWeb,” jjgl/index.html”）；若myWeb为null与方法（1）同。

（3）public URL（String protocol, String host, String fileName）。

这个构造方法中指定了协议名“protocol”、主机名“host”、文件名“fileName”，端口使用缺省值。如：

URL myurlhost=new URL(“http”,www.tsinghua.edu.on,”index.html”)。

（4）public URL（String protocol, String host,int port, String fileName）。

该构造方法与(3)构造方法相比较多了一个端口号“port”。如：

URL myurl port =new URL(“http”,www.tsinghua.edu.on, 80,”index.html”);

2．URL类中的常用方法：

```
//创建一个URL对象
    URL url = new URL("http://www.baidu.com");
    URL url1 = new URL(url, "/index.html?usrname=lqq#test");
//url的常用方法
    System.out.println("URL主机名称: "+url1.getHost());
    System.out.println("URL协议: "+url1.getProtocol());
    System.out.println("URL端口: "+url1.getPort());
    System.out.println("URL文件路径:"+url1.getPath());
    System.out.println("URL查询字符串: "+url1.getQuery());
    System.out.println("URL相对路径: "+url1.getRef());
    System.out.println("URL文件名: "+url1.getFile());
```

3．通过 URL 获取 html 文件内容：

```
//创建一个URL实例
URL url = new URL("http://www.baidu.com");
//通过URL对象的openstream方法获取一个InputStream对象
InputStream is = url.openStream();
InputStreamReader isr = new InputStreamReader(is, "utf-8");
BufferedReader br = new BufferedReader(isr);

File bd = new File(".../baidu.html");
PrintWriter pw = new PrintWriter(bd);
//按行读取
String s = br.readLine();
while(s!=null){
    pw.print(s);
    pw.flush();
    s=br.readLine();
}
//关闭相关资源
pw.close();
br.close();
isr.close();
is.close();
```

9.1.3 Socket编程

当两个程序需要通信时，可以通过使用 Socket 类建立套接字对象并连接在一起。网络上的两个程序通过一个双向的通信连接来实现数据的交换,这个双向链路的一端称为一个套接字（Socket）。Socket 通常用来实现客户方和服务方的连接。Socket 是 TCP/IP 协议的一个非常流行的方法，一个 Socket 由一个 IP 地址和一个端口确定。

套接字连接就是客户端的套接字对象和服务器端的套接字对象通过输入/输出流连接在一起，Socket 所要完成的通信就是基于连接的通信，建立连接所需的程序分别运行在客户端和服务器端。

首先，客户端发出对服务器端的连接申请，服务端程序不断监听端口，一旦有客户端发出的要连接某个端口的请求，就将 Socket 连接到此端口上，于是服务器和客户程序之间就建立了一个专用的虚拟连接。

建立连接后，客户程序可以向 Socket 写入请求，服务端程序开始处理请求并把处理结果再通过 Socket 返回给客户端，完成通过虚拟通道的数据通信。

通信结束后，将所建立的虚拟连接拆除。

Socket 通信过程如图 9-2 所示。

Java 在包 Java.net 中提供了两个类 Socket 和 ServerSocket，分别用来表示双向连接的客户端和服务端。

（1）客户建立连接到服务器的套接字对象 Socket 类

客户端的程序是使用 Socket 类建立与服务器套接字的连接。 Socket 类的构造方法如表 9-1 所示。

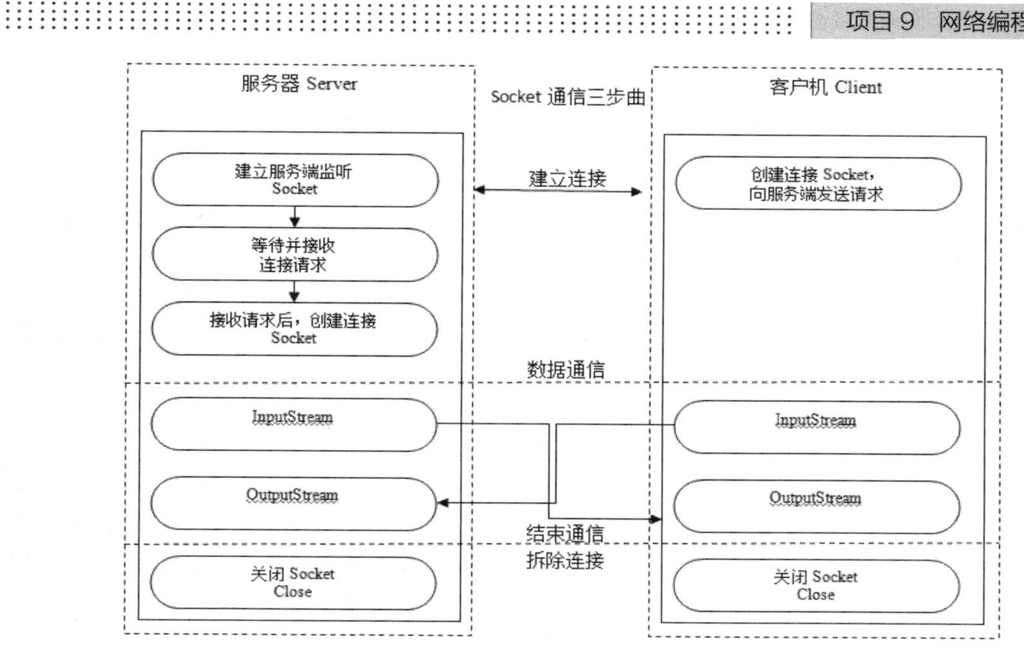

图 9-2　Socket 通信步骤示意图

表 9-1　Socket 类的主要构造方法

构造方法	功能说明
Socket（InetAddress addr，int port）	使用指定 IP 地址和端口创建一个 Socket 对象
Socket（InetAddress addr,int port,boolean stream）	使用指定地址和端口创建 Socket 对象,设置流式通信方式
Socket（String　host，int port）	使用指定主机和端口创建一个 Socket 对象
Socket（String　host,int port,boolean stream）	使用指定主机和端口创建 Socket 对象,设置流式通信方式

参数 host 是服务器的 IP 地址。port 是一个端口号。

建立时可能发生 IOException 异常,因此应像下面那样建立与服务器的套接字连接。

```
try { Socket clientsocket=new Socket("172.158.26.200",1880);}
catch(IOException e){}
```

当套接字连接 clientsocket 建立后,你可以想象一条通信“线路”已经建立起来。clientsocket 可以使用方法 getInputStream()获得一个输入流，然后用这个输入流读取服务器放入“线路”中的信息（但不能读取自己放入“线路”中的信息，就像打电话时，我们只能听到对方的声音一样）。clientsocket 还可以使用方法 getoutputStream()获得一个输出流，然后用这个输出流将信息写入“线路”中。

（2）负责等待客户端请求的套接字 ServerSocket

客户负责建立到服务器的套接字连接,服务器必须建立一个等待接收客户的套接字对象——ServerSocket。ServerSocket 类的构造方法如表 9-2 所示。

表 9-2　ServerSocket 类的主要构造方法

构造方法	功能说明
ServerSocket（int port）	在指定的端口创建一个 ServerSocket 对象
ServerSocket(int port,int count)	在指定的端口创建 ServerSocket 对象并指定服务器所能支持的最大连接数

port 是一个端口号。port 必须和客户呼叫的端口号相同。

建立时可能发生 IOException 异常,因此应像下面那样建立服务器的套接字。

```
try { Socket  server_socket=new ServerSocket(1880);}
catch(IOException e){}
```

当服务器的套接字 server_socket 建立后，可用 accept()方法接收客户套接字 clientsocket 的连接请求。

```
server_socket.accept();
```

接收客户套接字的过程也可能发生 IOException 异常,因此应像下面那样建立接收客户套接字。

```
try { Socket  sc=  server_socket.accept();}
catch(IOException e)
```

当收到客户的 clientsocket 套接字后，把它放到一个已声明的 Socket 对象 sc 中，那么 sc 就是 clientsocket，这样服务器端的 sc 就可以使用方法 getoutputStream()获得一个输出流，然后用这个输出流将信息写入“线路”中，发送给客户端。可以使用方法 getInputStream()获得一个输入流，然后用这个输入流读取客户放入“线路”中的信息。下面通过一个简单的例子说明上面的概念。

① 客户端程序

```
package com.guantong.chapter09;
/**
* client.java
* 客户端程序的实现
*/
import java.io.*;
import java.net.*;
public class client {
public static void main(String args[]){
   String str=null;
   Socket clientsocket;
   DataInputStream in=null;
   DataOutputStream out=null;
   try {   clientsocket=new Socket("localhost",4880);
    in=new DataInputStream(clientsocket.getInputStream());
    out=new DataOutputStream(clientsocket.getOutputStream());
    out.writeUTF("你好,我是客户机"); //通过 out 向"线路"写信息
    while (true) {
          str=in.readUTF(); //通过使用 in 读取服务器放入"线路"里的信息
          if (str!=null) break;
   }
    out.close();
    in.close();
    clientsocket.close();
    }catch (Exception e) {System.out.println("无法连接");}
  }
}
```

② 服务器端程序

```
package com.guantong.chapter09;
/**
* server.java
* 服务端程序的实现
*/
import java.io.*;
import java.net.*;
public class server {
    public static void main(String args[]){
        String str=null;
        ServerSocket server_Socket=null;
        Socket c_socket;
        DataInputStream in=null;
        DataOutputStream out=null;
        try {  server_Socket=new ServerSocket(4880);
        }catch (Exception e) {System.out.println("无法创建");}
        try{   c_socket=server_Socket.accept();
               in=new
        DataInputStream(c_socket.getInputStream());
               out=new
        DataOutputStream(c_socket.getOutputStream());
              while (true) {
              str=in.readUTF();//通过使用in读取客户放入"线路"里的信息
              if (str!=null) break;
      }
       out.writeUTF("你好,我是服务器"); //通过out向"线路"写信息
       out.close();
       in.close();
       c_socket.close();
    }catch (Exception e) {System.out.println("出现错误");}
 }
}
```

为了方便，我们建立套接字时，使用的服务器地址是localhost，它代表本地机IP地址。

✧ 编码实施

1. 服务器端程序代码

```
package com.guantong.chapter09;
/**
* TcpServer.java
* 服务端程序的实现
*/
import java.io.*;
import java.net.*;
  public class TcpServer {
     public static void main(String args[]){
     ServerSocket serversk;
```

```
        Socket sck;
        DataInputStream in;
        DataOutputStream out;
        InetAddress cltIP;
        String str="";
        try{
      InetAddress cltIP1=InetAddress.getLocalHost();
      serversk=new ServerSocket(9000);
         System.out.println("等待客户机的连接...");
         sck=serversk.accept(); //服务器等待客户的连接
         in=new DataInputStream(sck.getInputStream());
         out=new
      DataOutputStream(sck.getOutputStream());
         cltIP=sck.getInetAddress();
         System.out.println("客户机的 IP 地为: "+cltIP);
         out.writeUTF("欢迎客户机的访问...");
         str=in.readUTF();
         System.out.println(str);
         while (!str.equals("quit")){
              System.out.println("客户机: "+str);
              str=in.readUTF();
                }
       System.out.println("客户机: "+cltIP+"断开连接");
         out.close();
         in.close();
         sck.close();
         }catch (Exception e)
        {System.out.println(e.getMessage());}
       }
  }
```

2．客户端程序代码

```
  package com.guantong.chapter09;
  /**
  * TcpServer.java
  * 客户端程序的实现
  */
  import java.io.*;
  import java.net.*;
  public class TcpClient {
     public static void main(String args[]){
        Socket sck;
        DataInputStream in;
        DataOutputStream out;
        String str=null;
        try {   sck=new Socket("localhost",9000);
         System.out.println("正在连接到服务器 localhost...");
         in=new DataInputStream(sck.getInputStream());
         out=new
```

```
            DataOutputStream(sck.getOutputStream());
            str=in.readUTF();
            System.out.println("服务器: "+str);
            byte keychar[]=new byte[30];
            System.in.read(keychar);
            String msg=new String(keychar,0);
            msg.trim();
            while (!msg.equals("quit")){
                        out.writeUTF(msg);
               System.in.read(keychar);
               msg=new String(keychar,0);
               msg.trim();
              }
            out.writeUTF(msg);
            out.close();
            in.close();
            sck.close();
            }catch (Exception e)
           {System.out.println(e.getMessage());}
          }
    }
```

运行效果如图 9-3 所示。

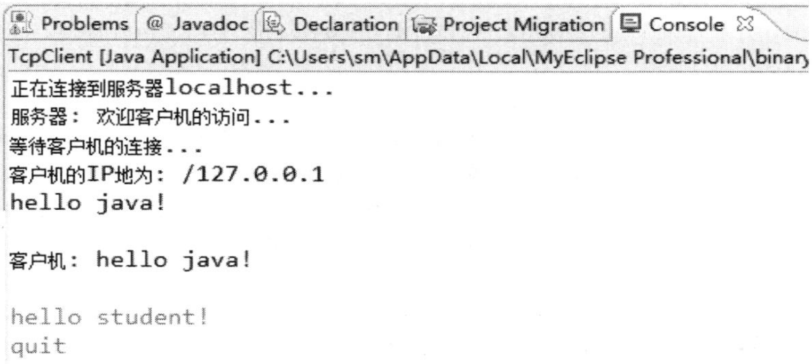

```
Problems  @ Javadoc  Declaration  Project Migration  Console
TcpClient [Java Application] C:\Users\sm\AppData\Local\MyEclipse Professional\binary
正在连接到服务器localhost...
服务器: 欢迎客户机的访问...
等待客户机的连接...
客户机的IP地为: /127.0.0.1
hello java!

客户机: hello java!

hello student!
quit
```

图 9-3 基于 TCP 的聊天程序

✧ 调试运行

1．服务器端程序 TcpServer 类，得到的程序内容框架如下。

```
public class TcpServer{
   public static void main(String args[]){
      ……  //详细实现参见“编码实施”代码
   }
}
```

2．创建客户端程序 TcpClient 类，得到的程序内容框架如下：

```
public class TcpClient{
   public static void main(String args[]){
```

```
        ……  //详细实现参见“编码实施”代码
    }
}
```

✧ 维护升级

本例中，服务器每次只能连接一个客户端，只有当连接的客户端中断后才能连接下一个客户端，如果要服务器同时处理多个客户端，应需要将服务器程序设计成多线程应用程序，修改的部分代码如下。

```
import java.io.*;import java.net.*;
import java.awt.*;import java.awt.event.*;
public class ClientTalkEx extends Frame implements ActionListener{
    Label label=new Label("客户机交谈内容：");
    Panel panel=new Panel();
    TextField txf=new TextField(12);
    TextArea txa=new TextArea();
    Socket client;InputStream in;OutputStream out;
public ClientTalkEx(){          //构造方法
    super("客户机 YU 交互聊天室");
    setSize(280,270);panel.add(label); panel.add(txf);
    txf.addActionListener(this);    //给文本框注册监听器
    add("North",panel);add("Center",txa);
addWindowListener(new WindowAdapter(){ //给框架注册监听器
    public void windowClosing(WindowEvent e){System.exit(0);}
    });show();
    try{
        client=new Socket(InetAddress.getLocalHost(),4000);//建立套接字获取信息
        txa.append("服务器名称是："+client.getInetAddress().getHostName()+"\n\n");
        in=client.getInputStream();out=client.getOutputStream();//获取输入输出流
        }catch (IOException ioe){}
    while(true){
        try{
            byte[] buf=new byte[256];
            in.read(buf);
            String str=new String(buf);
            txa.append("服务器畅谈："+str+"\n");
            }catch(IOException e){}}
    }
public void actionPerformed(ActionEvent e){
    try{
        String str=txf.getText();
        byte[] buf=str.getBytes();
        txf.setText(null);out.write(buf);
        txa.append("我 Client 说："+str+"\n");
        }catch(IOException iOE){} }
public static void main(String args[]){     //主方法
    new ClientTalkEx(); }
}
```

服务器端代码：

```
import java.io.*;import java.net.*;
import java.awt.*;import java.awt.event.*;
public class ServerTalkEx extends Frame implements ActionListener{
    Label label=new Label("服务器交谈内容：");
    Panel panel=new Panel();
    TextField txf=new TextField(12);
    TextArea txa=new TextArea();
    ServerSocket server;Socket client;
    InputStream in;OutputStream out;
public ServerTalkEx(){            //构造方法
    super("服务器YU交互聊天室");
    setSize(280,270);
    panel.add(label); panel.add(txf);
    txf.addActionListener(this);    //给文本框注册监听器
    add("North",panel);add("Center",txa);
    addWindowListener(new WindowAdapter(){//给框架注册监听器
        public void windowClosing(WindowEvent e){System.exit(0);}
    });show();
    try{
        server=new ServerSocket(4000);
        client=server.accept();      //从服务器套接字接收信息
        txa.append("客户机名称是："+client.getInetAddress().getHostName()+"\n\n");
        in=client.getInputStream();out=client.getOutputStream();}/
        catch (IOException ioe){}
    while(true){
        try {
            byte[] buf=new byte[256];
            in.read(buf);
            String str=new String(buf);
            txa.append("客户机述说："+str+"\n");}
        catch (IOException e){} }
    }
public void actionPerformed(ActionEvent e){
    try{
        String str=txf.getText();
        byte[] buf=str.getBytes();
        txf.setText(null);out.write(buf);
        txa.append("我Server说："+str+"\n");
        }catch (IOException ioe){}}
public static void main(String[] args){   //主方法
    new ServerTalkEx(); }
}
```

修改代码后，重新运行程序，程序运行的结果如图9-1所示。

任务 2 使用 UDP 协议的 Java 聊天室

✧ 需求分析

应用 Java 图形用户界面技术，编写一个基于 UDP 数据报协议的聊天程序。

1．需求描述

本任务是基于 UDP 的无连接的网络应用程序，首先要创建好如图 9-4 的用户界面。然后编写接收数据包的功能 UdpChat 类，编写每次“send”按钮就发送数据包到目的端主机以及显示出已发送的信息的 myMouseListener 类。

2．运行结果

运行结果如图 9-4 所示。

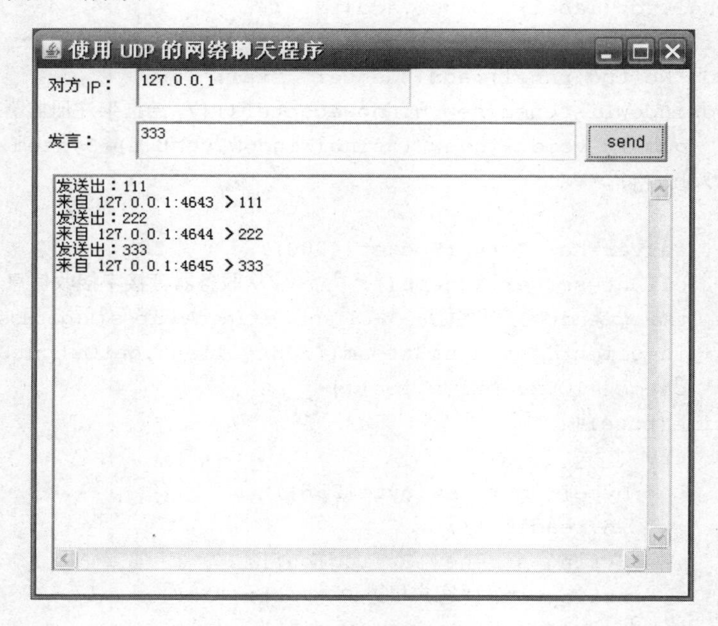

图 9-4 使用 UDP 的网络聊天程序

✧ 知识准备

1．基于 UDP 协议的网络编程

基于 TCP 的网络套接字（Socket），可以形象地比喻为打电话，一方呼叫，另一方负责监听，一旦建立了套接字连接，双方就可以互向通信了。

还有一种 UDP 通信传输方式，类似邮递信件，不是实时接收，当我们要求数据只要能以较快速地传输信息，并能容忍小的错误时，可以考虑一种基于 UDP（用户数据报协议）的网络通信传输方式。用户数据包协议是工作在传输层的面向无连接的协议，它的信息传递更快，但不提供可靠性保证。这种网络信息传输方式是数据在传输时，用户无法知道数据能否正确达到目的地主机，也不能确定数据到达目的地的顺序是否和发送的顺序相同。

UDP 协议的主要作用是将网络数据流量压缩成数据包的形式。在用 Java 实现 UDP 协议编程的过程中，需要用到两个套接字类：DatagramSocket 和 DatagramPacket；其中 Datagram Socket 是实现数据接收与发送的 Socket 实例；DatagramPacket 是实现数据封装实例，它将 Byte 数组、目标地址、目标端口等数据包装成报文或者将报文拆卸成 Byte 数组。

UDP 服务器要执行以下三步：

（1）创建一个 DatagramSocket 实例，指定本地端口号，并可以选择指定本地地址。此时，服务器已经准备好从任何客户端接收数据报文。

（2）使用 DatagramSocket 类的 receive()方法接收一个 DatagramPacket 实例。当 receive()方法返回时，数据报文就包含了客户端的地址与端口，这样我们就知道回复信息该发送到什么地方。

（3）使用 DatagramSocket 类的 send()和 receive()方法发送和接收 DatagramPacket 实例，进行通信。

2．InetAddress 类

Internet 上的主机有两种方式表示地址。

（1）域名：例如，www.sina.com 或 www.163.com。

（2）IP 地址：例如，202.104.35.210。

Java.net 包中的 InetAddress 类对象含有一个 Internet 主机地址的域名和 IP 地址：www.sina.com.cn/202.108.35.210。

我们可以使用 InetAddress 类的静态方法 getByName(s)来获得一个 InetAddress 对象，参数 s 为域名或 IP 地址。另外 InetAddress 类中还含有两个实例方法：

Public String getHostName()：获取 InetAddress 对象所含的域名。

Public String getHosAddress()：获取 InetAddress 对象所含的 IP 地址。

下面的示例展示了 InetAddress 对象的创建，以及如何获得 InetAddress 对象所含的域名和 IP 地址。

【例 9-1】 InetAddress 对象的创建示例。

```
            package com.guantong.chapter09;
    import java.net.*;
    public class getAddrIP {
            public static void main(String args[])
               {
             try{
                            InetAddress
addressA=InetAddress.getByName("www.sina.com.cn");
                        System.out.println(addressA.toString());
                         InetAddress
addressB=InetAddress.getByName("61.135.131.180");
                         //www.sohu.com
                         System.out.println(addressB.toString());
                }
                 catch(UnknownHostException e)      {
                 System.out.println("难以找到www.sina.com.cn,网络连接否?网址对否?");}
          }
    }
```

运行结果如图9-5所示。

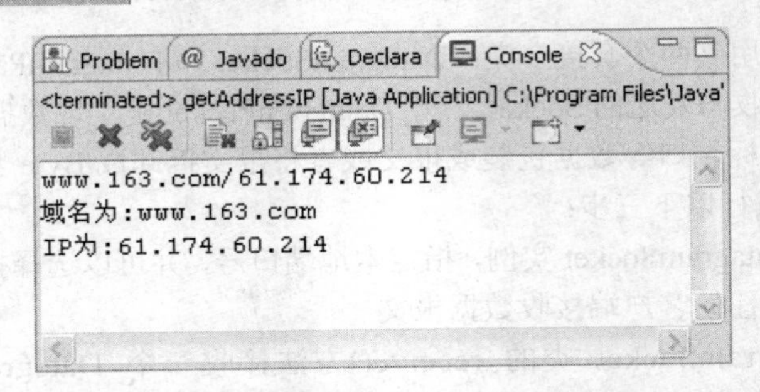

图 9-5 获得 InetAddress 对象所含的域名和 IP 地址

3. 基于 UDP 通信的基本模式

基于 UDP 通信的基本模式如下。

（1）发送数据包

① 首先用 DatagramPacket 类将数据打包，即用 DatagramPacket 类创建一个对象，称为数据包。用 DatagramPacket 的以下两个构造方法创建待发送的数据包：

```
DatagramPacket(byte data[ ],int length,InetAddress address,int port)
```

参数 data 中存放数据报数据，length 为数据报中数据的长度，address 表示数据报将发送到的目的地址。address 表示数据报将发送到的目的端口号。

```
DatagramPacket(byte data[ ],int offset,int length,InetAddress address,int port)
```

使用该构造方法创建的数据报对象含有数组 data 从 offset 开始指定长度的数据，该数据包将发送到地址是 address，端口号是 port 的主机上。例如：

```
byte data[ ]="近来你好吗".getByte();
InetAddress address= InetAddress.getByName(www.163.com);
DatagramPacket dp=new DatagramPacket(data,data.length,address,2000);
```

② 然后用 DatagramSocket 类的不带参数的构造方法创建一个对象，该对象负责发送数据包。

```
DatagramSocket ds=new DatagramSocket();
Ds.send(dp); //发送数据包
```

（2）接收数据包 用 DatagramSocket 类的另一种构造方法 DatagramSocket（int port）创建一个对象,其中的参数必须和待接收的数据包的端口相同，例如：如果发送方发送的数据包的端口是 6000

```
DatagramSocket data_in= DatagramSocket(6000);
```

则该对象 data_in 使用 receive(DatagramPacket pack)接收数据包，有一个数据包对数 pack，receive 把收到的数据包传递给该参数。因此我们必须预备一个数据包以便收取数据包，这时需使用 DatagramPacket 类的另外一个构造方法 DatagramPacket（byte data[ ],int length）创建一个数据包，用于接收数据包。例如：

```
byte data[ ] =new byte[200];
DatagramPacket  pack=new  DatagramPacket(byte data[ ],100);
data_in.reveive(pack);
```

该数据包 pack 接收数据的长度为 100 个字节。

✧ 编码实施

```
package com.guantong.chapter09;
```

```
import java.awt.*;
import java.net.*;
import java.awt.event.*;
public class UdpChat extends Frame implements Runnable{
  Label L1,L2;
  TextField text1,text2;
  Button B1;
  TextArea messageArea;
  public UdpChat()  {
    this.setLayout( null );
    //=========================================
    L1 = new Label("对方 IP: ");
    L1.setBounds(10,30,60,30);
    this.add(L1);
    L2 = new Label("发言: ");
    L2.setBounds(10,70,60,30);
    this.add(L2);
    text1 = new TextField("127.0.0.1", 20);
    text1.setBounds(75,30,200,30);
    this.add(text1);
    text2 = new TextField();
    text2.setBounds(75,70,320,30);
    this.add(text2);
    B1 = new Button("send");
    B1.setBounds(400,70,60,30);
    B1.addMouseListener( new myMouseListener() );
    this.add(B1);
    messageArea=new
    TextArea("",20,20,TextArea.SCROLLBARS_BOTH);
    messageArea.setBounds(15,110,450,300);
    this.add(messageArea);
    //=========================================
    this.addWindowListener(new WindowAdapter() {
      public void windowClosing(WindowEvent e) {
        System.exit(0);
      }
    });
    this.setTitle("使用 UDP 的网络聊天程序");
    this.setBounds(100,100,480,430);
    this.setVisible( true );
  }
  public void run()  {  //接收数据
     while( true ) { //持续接收送到本地端的信息
        byte[] buf = new byte[100];//预期最多可收 100 个 byte
        try{ DatagramSocket DS = new DatagramSocket( 2222 );
                                                //用 2222 port 收
/* 只管接收要送到本地端 2222 port 的数据包,
不必管该数据包是从远程的那个 port 送出。 */
```

```
DatagramPacket DP = new DatagramPacket( buf,buf.length );
                                        //将数据收到 buf 数组
        DS.receive( DP ); //接收数据包
        messageArea.append( "来自" + DP.getAddress().getHostAddress()
              + ":" + DP.getPort() + " >"
              + new String( buf ).trim()+"\n" );
        DS.close();
        Thread.sleep(200); //停 0.2 秒
      }
      catch(Exception excep){}
    }
  } // void run() end
  class myMouseListener extends MouseAdapter { //送数据
public void mouseClicked(MouseEvent e){
//每次 Click 按钮就发送信息到目的端主机
    String msg = text2.getText().trim();
    String ipStr = text1.getText().trim();
    try{   DatagramSocket DS = new DatagramSocket();

//以任一目前可用的 port 送
      DatagramPacket DP = new DatagramPacket( msg.getBytes(),
      msg.getBytes().length, InetAddress.getByName(ipStr), 2222 );
//送到远程的 2222 port
      DS.send( DP ); //送出数据包
      messageArea.append( "发送出: " + msg.trim() +"\n" );
                                        //给自己看的记录

      DS.close();
    }
    catch(Exception excep){}
  }
  }
  public static void main(String arg[])  {
UdpChat udpchat = new UdpChat();
   Thread threadObj = new Thread(udpchat);
   threadObj.start();//启动接收信息的线程
}
}
```

✧ 调试运行

创建基于 UDP 的网络聊天程序文件 UdpChat.java，得到的程序内容框架如下。

```
package com.guantong.chapter09;
import java.awt.*;
import java.net.*;
import java.awt.event.*;
public class UdpChat extends Frame implements Runnable{
  Label L1,L2;
  TextField text1,text2;
```

```
 Button B1;
 TextArea messageArea;
 public UdpChat(){
   ......        /参见详细实现代码
 }
 public void run()  {
   ......        //参见详细实现代码
         }
 class myMouseListener extends MouseAdapter{  //送数据
  public void mouseClicked(MouseEvent e) {
   ......         //参见详细实现代码
   }
   public static void main(String arg[]){
......         //参见详细实现代码
 }
}
```

✧ 维护升级

本例中，聊天的内容不能长期保存，可以在此界面中增加一个“保存记录”按钮，单击此按钮将聊天记录保存在 Note.txt 文件中。修改后的代码如下：

```
package com.guantong.chapter09;
import java.awt.*;
import java.net.*;
import java.awt.event.*;
import java.io.*;
public class UdpChat extends Frame implements Runnable{
  Label L1,L2;
  TextField text1,text2;
  Button B1,B2;
  TextArea messageArea;
   Public UdpChat()  {
    this.setLayout( null );
    //========================================
    L1 = new Label("对方 IP: ");
    L1.setBounds(10,30,60,30);
    this.add(L1);
    L2 = new Label("发言: ");
    L2.setBounds(10,70,60,30);
    this.add(L2);
    text1 = new TextField("127.0.0.1", 20);
    text1.setBounds(75,30,200,30);
    this.add(text1);
    text2 = new TextField();
    text2.setBounds(75,70,320,30);
    this.add(text2);
    B1 = new Button("send");
    B1.setBounds(400,70,60,30);
    B1.addMouseListener( new myMouseListener() );
```

```
    B2 = new Button("保存记录");
    B2.setBounds(400,35,60,30);
    B2.addMouseListener( new SaveMsg() );
    this.add(B1);
    this.add(B2);
    messageArea=new TextArea("",20,20,TextArea.SCROLLBARS_BOTH);
    messageArea.setBounds(15,110,450,300);
    this.add(messageArea);
    //========================================
    this.addWindowListener(new WindowAdapter()  {
     public void windowClosing(WindowEvent e) {
       System.exit(0);
     }
    });
    this.setTitle("使用 UDP 的网络聊天程序");
    this.setBounds(100,100,480,430);
    this.setVisible( true );
  }
  public void run()   {  //接收数据
    while( true ) { //持续接收送到本地端的信息
      byte[] buf = new byte[100];//预期最多可收 100 个 byte
      try {DatagramSocket DS = new DatagramSocket( 2222 );
                                          //用 2222 port 收
         /* 只管接收要送到本地端 2222 port 的数据包,
            不必管该数据包是从远程的哪个 port 送出。 */
        DatagramPacket DP = new DatagramPacket( buf,buf.length );
                                         //将数据收到 buf 数组
        DS.receive( DP ); //接收数据包
        messageArea.append( "来自" + DP.getAddress().getHostAddress()
                 + ":" + DP.getPort() + " >"
                 + new String( buf ).trim()+"\n" );
           //此处用 new String(DP.getData()).trim() 也一样
        DS.close();
        Thread.sleep(200); //停 0.2 秒
      }
      catch(Exception excep){}
    }
  }
  class myMouseListener extends MouseAdapter {  //送数据
   public void mouseClicked(MouseEvent e) { //每次 Click 按钮就发送信息到
目的端主机
     String msg = text2.getText().trim();
     String ipStr = text1.getText().trim();
     try {DatagramSocket DS = new DatagramSocket(); //以任一目前可用的port 送
       DatagramPacket DP
         = new DatagramPacket( msg.getBytes(),
                          msg.getBytes().length,
                          InetAddress.getByName(ipStr),
                          2222 ); //送到远程的 2222 port
```

```
            DS.send( DP ); //送出数据包
            messageArea.append( "发送出: " + msg.trim() +"\n" );//给自己看的记
录
            DS.close();
          }
          catch(Exception excep){}
        }
      }
    //聊天记录的保存 保存在当前位置下的"Note.txt"文件中
      class SaveMsg extends MouseAdapter  {
         String msg=null;
         String line =System.getProperty("line.separator");
         public void mouseClicked(MouseEvent e)  {
             try  {   msg=messageArea.getText();
                FileOutputStream Note=new FileOutputStream("Note.txt");
                messageArea.append("记录已经保存在 Note.txt");
                Note.write(msg.getBytes());
                messageArea.append(line);
                Note.close();
              }
             catch (IOException e1) {
                System.out.println("发送失败");
             }
         }
      }
      public static void main(String arg[])  {
    UdpChat udpChat = new UdpChat ();
         Thread threadObj = new Thread(udpChat);
         threadObj.start();//启动接收信息的线程
    }
    }
```

修改代码后，重新运行程序，程序运行的结果如图 9-6 所示。

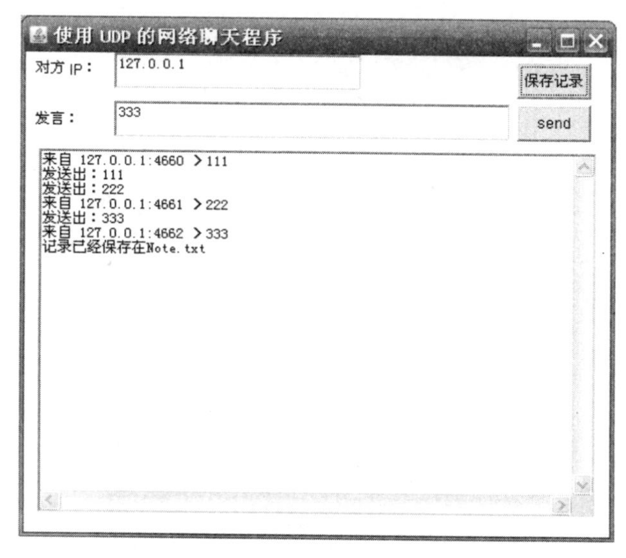

图 9-6　具有保存功能的 UDP 的网络聊天程序的运行结果

项目实训与练习

一、简答题

1．什么是TCP/IP协议？

2．TCP/IP有哪两种传输协议？各有什么特点？

3．什么是URL?

4．一个URL对象通常包含哪些信息？

5．java使用哪个组件来显示URL中的HTML文件？

6．JAVA网络编程中有哪些常用的类？

7．客户端的Socket对象和服务器端的Socket对象是怎样通信的？

8．基于UDP的通信和基于TCP的通信有什么不同？

9．一个完整的URL地址由哪几部分组成？

二、编程题

试完成用URL读取网络上的文本信息资源，并将其保存到指定文件中的编程。

项目实战——学生信息管理系统

项目目标

在熟练掌握JAVA GUI编程和JDBC编程的基础上，利用软件工程开发项目的步骤和设计思想，开发具有实际应用价值的学生信息管理系统。通过项目的学习，使同学们巩固所学知识点和主要技能点，为日后从事项目开发打下良好基础。

项目内容

本项目将详细介绍学生信息管理系统的实现过程，如系统的需求分析、概要设计、数据库设计、模块实现和系统测试等。

10.1 系统概述

学生信息管理系统主要用于学校学生信息管理，实现学生信息资源的系统化、科学化、规范化和自动化管理，其主要任务是用计算机对学生各种信息进行日常管理，如查询、修改、增加、删除，针对这些要求设计了学生信息管理系统。推行学校信息管理系统的应用是进一步推进学生学籍管理规范化、电子化的重要举措。

学生信息管理对于学校的管理者来说至关重要，学生信息是高等学校非常重要的一项数据资源，是一个教育单位不可缺少的一部分。特别是近几年来，国家政策的调整，我国高等院校大规模的扩招，给高等院校的教学管理、学生管理、后勤管理等方面都带来不少新的课题。其包含的数据量大，涉及的人员面广，而且需要及时更新，故较为复杂，难以单纯地信赖人工管理，而且传统的人工管理方式既不易于规范化，管理效率也不高。随着科学技术的不断提高，计算机科学与技术日渐成熟，计算机应用的普及已进入人类社会生活的各个领域，并发挥着越来越重要的作用。传统手工信息的管理模式必然被以计算机为物质基础的信息管理方法所取代。

学生信息管理系统针对在校学生信息的特点，以及管理中实际需要而设计。有效地实现学生信息管理的信息化，高效率、规范化地管理大量的学生信息，并避免人为操作造成的错误和不规范行为，本系统便是针对这样的背景开发而成，同时学生信息管理系统，也可以作为一种教学资源，供学生和教师在日常的教学活动学习和讲解知识时使用的综合案例。

10.2 需求分析

学生信息管理系统，主要是管理学生基本情况信息，对于学生信息的管理应该包括对学生信息的添加、删除、修改、查询操作。同时在项目开发的过程中，根据使用本系统的对象，设置了不同身份的用户。如：管理员、教师、学生。不同用户对学生信息管理系统的使用权限有所不同。例如管理员可以对学生信息进行添加、删除、修改、查询操作；教师只有学生信息的浏览权限；学生用户也只具有对学生信息的浏览权限。同时根据日常需要，系统对不同身份的用户也实现了管理操作，如管理员可以查询系统所有用户的信息，可以添加使用系统的用户身份，可以删除使用本系统的人员信息。教师可以修改自己登录系统时使用的密码。学生可以修改登录时使用的原始密码。系统使用完成后，可以退出系统的功能。

10.2.1 本系统开发过程中使用的环境

1．硬件环境

处理器 Pentium3 以上；内存 512MB 以上；硬盘 40GB 以上。

2．软件环境

（1）操作系统：Windows XP 或 Windows 7 及以上版本的操作系统。

（2）数据库服务器：Microsoft SQL Server2005。

（3）开发工具：jdk1.5 版本以上;eclipse3.0 以上版本。

10.2.2 概要设计

1．模块设计

经过需求分析得出本系统应该具有如下功能（见图 10-1）。

（1）用户权限判断　登录系统时，不同身份的用户所具有的权限不同，如管理员具有用户管理功能，包括对用户的添加、删除、修改、查询功能；教师具有对自己原始密码修改功能；查询学生信息的功能。学生对自己原始密码修改功能，查询学生信息功能。

（2）系统管理　根据用户身份的不同可以对用户进行用户的添加、删除、修改、查询操作。

（3）学生信息管理　根据用户身份的不同可以对学生信息进行添加、删除、修改、查询操作。

（4）重新登录　用户可以重新登录本系统。

（5）系统退出　使用者可以退出本系统。

2．数据库设计

本系统所需要的永久性数据都存储在 Microsoft SQL Server2005 数据库系统中，在 SQL Server2005 数据库系统中建立一个名字为 StuInfo 的数据库，该数据库里包括用户身份表 UserInfo、学生信息表 StudentInfo、教师信息表 TeacherInfo。

（1）用户身份表UserInfo　用户身份表UserInfo，用来存储使用本系统的用户信息，如登录系统时的用户名和密码以及用户权限。表结构如表10-1所示。

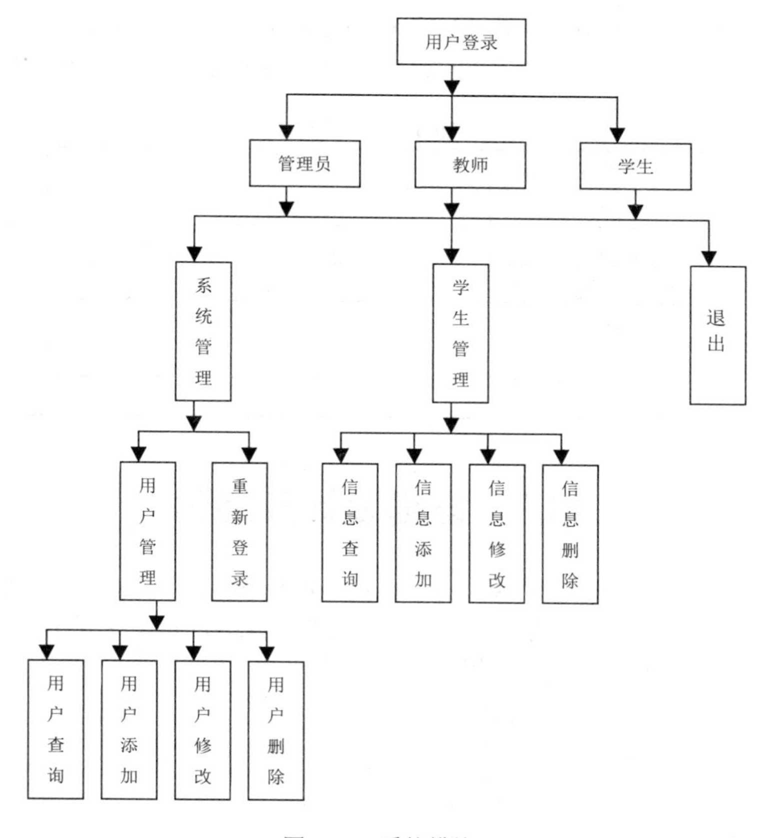

图 10-1 系统模块

表 10-1 用户信息表

字段名	数据类型	长度	含义	备注
Userid	varchar	10	教师 id	主键
Userpwd	varchar	10	姓名	
Userlevel	varchar	10	性别	

（2）学生信息表 StudentInfo 学生信息表 StudentInof，用来存储学生的基本信息。包括学生的学号、姓名、性别、出生日期、班级、学生电话、学生家庭住址。表结构如表 10-2 所示。

表 10-2 学生信息表

字段名	数据类型	长度	含义	备注
Sno	varchar	10	学号	主键
Sname	varchar	20	学生姓名	
Sex	varchar	2	性别	
Birthday	datetime		出生日期	
Classid	varchar	10	班级 id	
Tel	varchar	20	电话	
Adress	varchar	100	地址	

（3）教师信息表 TeacherInfo　教师信息表 TeacherInfo，用来存储教师的基本信息。包括教师的姓名、性别、办公室电话、家庭住址、出生日期。表结构如表 10-3 所示。

表 10-3　教师信息表

字段名	数据类型	长度	含义	备注
Teaid	varchar	10	教师 id	主键
Teaname	varchar	20	姓名	
Teasex	varchar	2	性别	
Teloffice	varchar	20	电话	
Address	varchar	100	住址	
TeaBirthday	datetime		生日	

10.3　详细设计及编码

打开 eclipse，创建一个名字叫 StudentInfo 的 JAVA Project，在该工程下引入数据库链接时所用到的 sqljdbc.jar 文件，使用 JDBC 链接 SQL Server2005 数据库系统时用到的驱动类就封装在此 jar 包中。该项目的组织结构所图 10-2 所示。

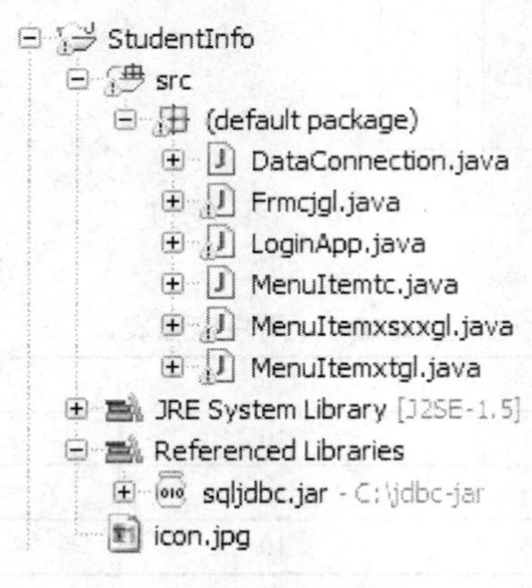

图 10-2　项目结构图

1．数据库链接类 DataConnection

在实现各个模块的功能之前，需要实现通用模块类的设计。为了减少代码的重复编写，在实现本系统的过程中把经常使用到的数据库链接操作封装成 DataConnection 类。通过对数据库链接类的封装可以使程序员在编写其他模块时不用重写该部分的代码，并且在对本系统更换数据库系统时，可以只修改数据库链接类，其他模块可以不用做任何改动。

DataConnection 类实现的代码如下。

```
import java.sql.Connection;
import java.sql.DriverManager;
import java.sql.Statement;
public class DataConnection {//数据库链接类
    public Statement stmt()
    {
```

```
            try {
                String name="sa";
                String pwd="123456";
                String
url="jdbc:sqlserver://localhost:1433;DataBaseName=StuInfo";

    Class.forName("com.microsoft.sqlserver.jdbc.SQLServerDriver");
                Connection con = DriverManager.getConnection(url,name,pwd);
                st = con.createStatement();
                } catch (Exception e)
                    {e.printStackTrace();}
        return st;
        }
        Statement st ;
    }
```

2. 登录界面类 LoginApp

LoginApp 类实现了用户登录本系统时的登录界面设计。如图 10-3 所示。

图 10-3　登录界面

实现代码如下：

```
    import java.awt.*;
    import java.awt.event.*;
    import java.sql.*;
    import javax.swing.*;
    public class LoginApp {//登录界面类
        public static String userid;
        public static String userpwd;
        public static String userlevel;//用户级别
        static Frmcjgl winmain;
        final JFrame app=new JFrame("登录界面");
        JLabel label0=new JLabel(new ImageIcon("icon.jpg"));
        JLabel label1=new JLabel("登录帐号");
        JLabel label2=new JLabel("登录密码");
        JButton button1=new JButton("登录");
        JButton button2=new JButton("取消");
        final JTextField textid=new JTextField(10);
        final JPasswordField textpwd=new JPasswordField(10);
```

```
    Container c=app.getContentPane();
    DataConnection con=new DataConnection();//创建数据库链接类的对象
  public LoginApp()
  {
    app.setDefaultCloseOperation(JFrame.EXIT_ON_CLOSE);
    app.setSize(500,270);
    app.setLocation(380, 270);
    app.setResizable(false);
    c.setLayout(new FlowLayout());//组件部局
    c.add(label0);//添加组件
    c.add(label1);
    c.add(textid);
    c.add(label2);
    c.add(textpwd);
    c.add(button1);
    c.add(button2);
    app.setVisible(true);
    button1.addMouseListener(new MouseAdapter()//用内部类实现鼠标单击事件，
该事件用来判断用户登录时用户名和密码是否存在。
    {
    public void mouseClicked(MouseEvent e)
        {
        if              (textid.getText().equals("")              ||
textpwd.getText().equals(""))
        JOptionPane.showMessageDialog(null, "请输入用户名或密码");
        else{
            userid=textid.getText();
            userpwd=textpwd.getText();
            try {
  String s_sql="SELECT Userpwd,Userlevel FROM UserInfo WHERE Userid='"+userid+"'";
                System.out.println(s_sql);
                ResultSet rs=con.stmt().executeQuery(s_sql);
                rs.next();
                if (rs.getRow()==0 ||
                        (!userpwd.equals(rs.getString(1)) ) )
        JOptionPane.showMessageDialog(null, "用户名或密码错误");
                else {
                    //进入成绩管理系统
                    userlevel=rs.getString(2);
                    //System.out.println(userlevel);
                    app.dispose();
                  winmain=new Frmcjgl();//登录成功后，打开主界面。
                    }
                con.stmt().close();
                } catch (SQLException e1)

                  e1.printStackTrace();
```

```
                    }
                }
            }
        });
        button2.addMouseListener(new MouseAdapter()//取消登录
        {
            public void mouseClicked(MouseEvent e)
            {
                textid.setText("");
                textpwd.setText("");
            }
        }
        );
    }
    public static void main(String[] args) {
        LoginApp login=new LoginApp();
    }
    }
```

3．主界面类 Frmcjgl

Frmcjgl 类实现了学生信息管理系统的主操作界面的设计，如图 10-4 所示。

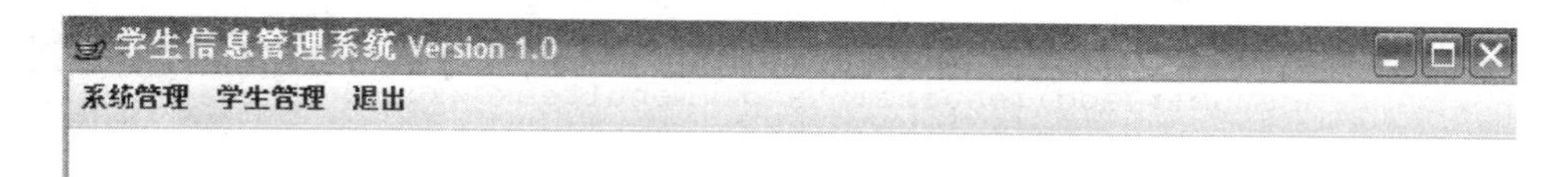

图 10-4　系统主界面

实现代码如下：

```
import java.awt.Container;
import javax.swing.*;
public class Frmcjgl {
    static JFrame app=new JFrame("学生信息管理系统 Version 1.0");
    private String userlevel;
    //实例化各个菜单项
    MenuItemxtgl menuitemxtgl=new MenuItemxtgl();//系统管理菜单
    MenuItemxsxxgl menuitemxsxxgl=new MenuItemxsxxgl();//学生信息管理菜单
    MenuItemtc menuitemtc=new MenuItemtc();//退出菜单
    public static JDesktopPane desktop=new JDesktopPane();
    public void setlevel()//判断登录用户权限的方法
    {
        if (userlevel.equals("学生"))  //设置学生权限
        {
            //系统管理
            menuitemxtgl.menuItemyhcx.setEnabled(false);//用户查询
            menuitemxtgl.menuItemyhtj.setEnabled(false);//用户添加
            menuitemxtgl.menuItemyhsc.setEnabled(false);//用户删除
```

```
        menuitemxsxxgl.menuItemxsxxtj.setEnabled(false);
        menuitemxsxxgl.menuItemxsxxxg.setEnabled(false);
        menuitemxsxxgl.menuItemxsxxsc.setEnabled(false);
    }
    else if (userlevel.equals("教师")) //设置教师权限
    {
        //系统管理
        menuitemxtgl.menuItemyhcx.setEnabled(false);
        menuitemxtgl.menuItemyhtj.setEnabled(false);
        menuitemxtgl.menuItemyhsc.setEnabled(false);
        //学生管理
        menuitemxsxxgl.menuItemxsxxtj.setEnabled(false);
        menuitemxsxxgl.menuItemxsxxxg.setEnabled(false);
        menuitemxsxxgl.menuItemxsxxsc.setEnabled(false);
    }
   }
   public void addmenu()//设置菜单栏
   {
    JMenuBar menubar=new JMenuBar();
    app.setJMenuBar(menubar);
    menubar.add(menuitemxtgl.menuItemxtgl);
    //学生管理
    menubar.add(menuitemxsxxgl.menuItemxsxxgl);
    //退出
    menubar.add(menuitemtc.menuItemtc);
   }
   public Frmcjgl(){
    app.setDefaultCloseOperation(JFrame.EXIT_ON_CLOSE);
    app.setSize(700,700);
    app.setLocation(180, 30);
    userlevel=LoginApp.userlevel;
    System.out.println(userlevel);
    setlevel();//设置菜单可选权限
    addmenu();//添加各个菜单
    //加入多文档窗口
    Container c=app.getContentPane();
   c.add(desktop);
    app.setVisible(true);
  }
}
```

4．系统管理类 MenuItemxtgl

MenuItemxtgl 类实现了系统管理菜单及子菜单用户管理、重新登录功能的设计。其中用户管理子菜单又包括用户的查询功能（如图 10-5 所示）、添加功能（如图 10-6 所示）、修改功能（如图 10-7 所示）和删除功能（如图 10-8 所示）。

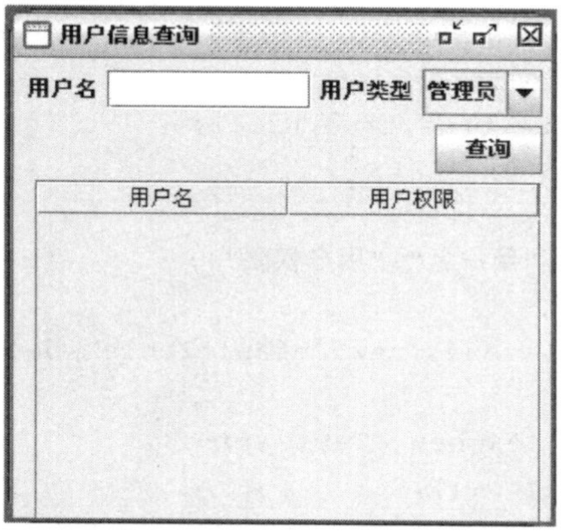

图 10-5　用户信息查询

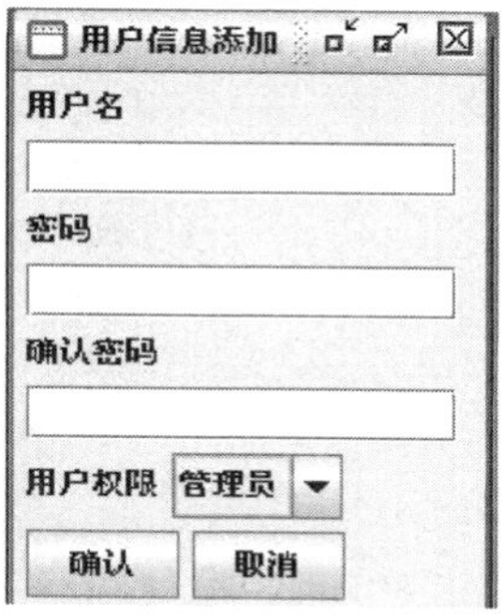

图 10-6　用户信息添加

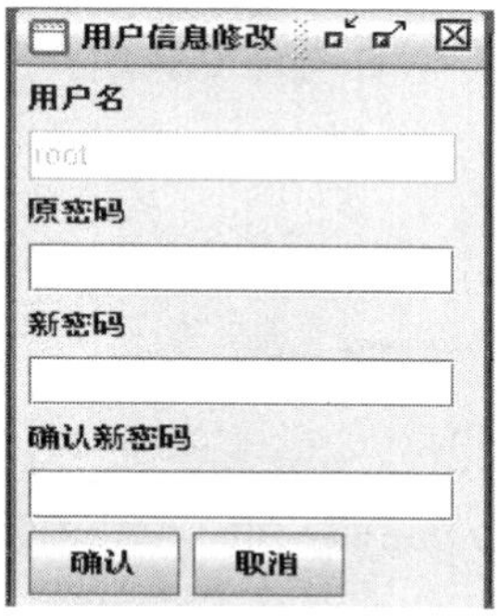

图 10-7　用户信息修改

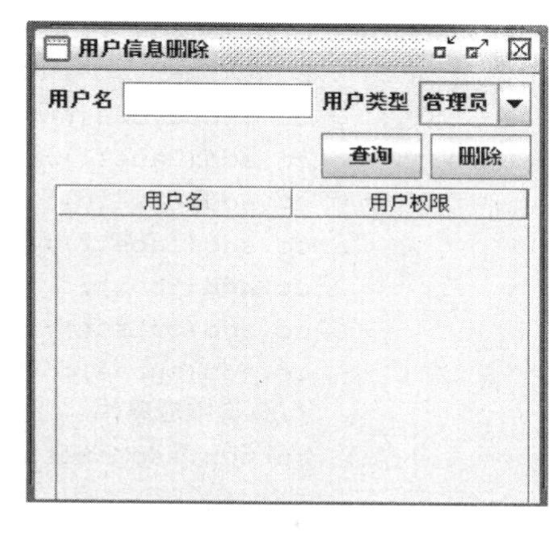

图 10-8　用户信息删除

实现代码如下：

```
import java.awt.*;
import java.awt.event.*;
import java.sql.*;
import java.util.Vector;
import javax.swing.*;
import javax.swing.table.*;
public class MenuItemxtgl {
    public void action_yhcx()//用户查询菜单响应
    {
        menuItemyhcx.addActionListener(new ActionListener()
        {
        public void actionPerformed(ActionEvent e)
        {
            String[] s={"管理员","教师","学生"};
            JButton button=new JButton("查询");
            JLabel label1=new JLabel("用户名");
            JLabel label2=new JLabel("用户类型");
            final JTextField textid=new JTextField(10);
```

```
            final JComboBox jbox=new JComboBox(s);
                final JInternalFrame internalframe=
new JInternalFrame("用户信息查询",true,true,true,true);
            internalframe.setSize(310, 300);
            //加入 JTable
            Object[] column_names={"用户名","用户权限"};
            Object[][] data={};
            final DefaultTableModel model=new DefaultTableModel
            (data,column_names);
                final JTable tableview=new JTable();
                tableview.setModel(model);
                tableview.setPreferredScrollableViewportSize(new Dimension
                (280,200));
            ableview.setAutoResizeMode(JTable.AUTO_RESIZE_SUBSEQUENT_COLUMNS);
                JScrollPane spane=new JScrollPane(tableview);
            Container cc=internalframe.getContentPane();
            cc.setLayout(new FlowLayout(FlowLayout.RIGHT));
            cc.add(label1);
            cc.add(textid);
            cc.add(label2);
            cc.add(jbox);
            cc.add(button);
            cc.add(spane);
            //查询响应事件
            button.addMouseListener(new MouseAdapter()
                {
                    public void mouseClicked(MouseEvent e)
                        {
                        try {
                            //用户名域为空
                            if (textid.getText().equals(""))
                                {
String s_sql="SELECT Userid,Userlevel FROM UserInfo WHERE Userlevel
='"+jbox.getSelectedItem()+"'";
            System.out.println(s_sql);
            ResultSet rs=con.stmt().executeQuery(s_sql);
                rs.next();
            //返回值为空
                if (rs.getRow()==0)
JOptionPane.showMessageDialog(null,"没有权限为"+jbox.getSelectedItem()+"的记录");
            //不为空，输出到 JTable 上
             else
            {
             //删除之前显示的行
            System.out.println(model.getRowCount());
             int count=model.getRowCount();
             for(int i=0;i<count;i++)
```

```
                model.removeRow(0);
                 do
                {
                    Vector vs=new Vector();
                     for (int i=1;i<3;i++)
                 vs.add(rs.getObject(i));
                     model.addRow(vs);
                }while (rs.next());
                    }
                }
                    //用户名域不为空
                else{
          String s_sql="SELECT Userid,Userlevel FROM UserInfo WHERE Userid='"
           +textid.getText()+"' and UserLevel='"+jbox.getSelectedItem()+"'";
            System.out.println(s_sql);
            ResultSet rs=con.stmt().executeQuery(s_sql);
            rs.next();
            //返回值为空
             if (rs.getRow()==0)
JOptionPane.showMessageDialog(null, "没有用户名为"+textid.getText()+"的记录");
              //不为空，输出到 JTable 上
             else
             {
             //删除之前显示的行
             System.out.println(model.getRowCount());
             int count=model.getRowCount();
             for(int i=0;i<count;i++)
             model.removeRow(0);
              do
              {
              Vector vs=new Vector();
              for (int i=1;i<3;i++)
              vs.add(rs.getObject(i));
              model.addRow(vs);
               }while (rs.next());
              }}
         } catch (SQLException e1)
         {e1.printStackTrace();}
         }
          });
         Frmcjgl.desktop.add(internalframe);
         internalframe.setVisible(true);
          }
         }
         );
    }
    public void action_yhtj()//用户信息添加菜单响应
```

```
    {
        menuItemyhtj.addActionListener(new ActionListener()
        {
        public void actionPerformed(ActionEvent e)
        {
    JLabel[] label={new JLabel("用户名"),new JLabel("密码"),
new JLabel("确认密码"),new JLabel("用户权限")};
            JButton[] button={new JButton("确认"),new JButton("取消")};
            final JTextField textid=new JTextField(15);
            final JPasswordField textpwd1=new JPasswordField(15);
            final JPasswordField textpwd2=new JPasswordField(15);
            String[] s={"管理员","学生","教师"};
            final JComboBox jbox=new JComboBox(s);
final JInternalFrame internalframe=
new JInternalFrame("用户信息添加",true,true,true,true);
            internalframe.setSize(195, 280);
            Container cc=internalframe.getContentPane();
            cc.setLayout(new FlowLayout(FlowLayout.LEFT));
            cc.add(label[0]);
            cc.add(textid);
            cc.add(label[1]);
            cc.add(textpwd1);
            cc.add(label[2]);
            cc.add(textpwd2);
            cc.add(label[3]);
            cc.add(jbox);
            cc.add(button[0]);
            cc.add(button[1]);
            Frmcjgl.desktop.add(internalframe);
            //点击确定按钮
            button[0].addMouseListener(new MouseAdapter()
            {
                public void mouseClicked(MouseEvent e)
                {
    if(textid.getText().equals("")||textpwd1.getText().equals("")
    ||textpwd2.getText().equals(""))
        JOptionPane.showMessageDialog(null, "信息请输入完全");
          else if (!textpwd1.getText().equals(textpwd2.getText()))
                        {
            JOptionPane.showMessageDialog(null, "两次密码不同，请重新输入");
            textpwd1.setText("");
            extpwd2.setText("");
            }
              else //信息正确进入数据库添加操作
                    {
                    try
                {
```

```
//判断用户名是否存在
String s_sql="SELECT Userlevel FROM UserInfo WHERE Userid='"
        +textid.getText()+"'";
System.out.println(s_sql);
ResultSet rs=con.stmt().executeQuery(s_sql);
rs.next();
if(rs.getRow()!=0)
JOptionPane.showMessageDialog(null,"用户名已存在");
else{
s_sql="INSERT INTO UserInfo(Userid,Userpwd,Userlevel) VALUES("
"'"+textid.getText()+"','"+textpwd1.getText()+"',
'"+jbox.getSelectedItem()+"')";
 System.out.println(s_sql);
int count=con.stmt().executeUpdate(s_sql);
 if (count==1)
JOptionPane.showMessageDialog(null,"添加成功");
else JOptionPane.showMessageDialog(null, "添加失败");
}
} catch (Exception e1)
{e1.printStackTrace();}
    }
   }
  }
 );
 //点击取消按钮
button[1].addMouseListener(new MouseAdapter()
  {
   public void mouseClicked(MouseEvent e)
   {
    internalframe.dispose();
    }
  });
internalframe.setVisible(true);
  }
    });
}
public void action_yhxg()//用户信息修改菜单响应
{
menuItemyhxg.addActionListener(new ActionListener()
  {
   public void actionPerformed(ActionEvent e)
   {
JLabel[] label={new JLabel("用户名"),new JLabel("原密码"),
new JLabel("新密码"),new JLabel("确认新密码")};
    JButton[] button={new JButton("确认"),new JButton("取消")};
    final JTextField textid=new JTextField(15);
    final JPasswordField textpwd_before=new JPasswordField(15);
```

```
        final JPasswordField textpwd_new1=new JPasswordField(15);
        final JPasswordField textpwd_new2=new JPasswordField(15);
        final JInternalFrame internalframe=
new JInternalFrame("用户信息修改",true,true,true,true);
            internalframe.setSize(195, 280);
            Container cc=internalframe.getContentPane();
            cc.setLayout(new FlowLayout(FlowLayout.LEFT));
            Frmcjgl.desktop.add(internalframe);
            cc.add(label[0]);
            cc.add(textid);
            cc.add(label[1]);
            cc.add(textpwd_before);
            cc.add(label[2]);
            cc.add(textpwd_new1);
            cc.add(label[3]);
            cc.add(textpwd_new2);
            cc.add(button[0]);
            cc.add(button[1]);
            //设置用户名不可更改
            textid.setText(LoginApp.userid);
            textid.setEnabled(false);

            //确定按钮响应
            button[0].addMouseListener(new MouseAdapter()
            {
                public void mouseClicked(MouseEvent e)
                {
            if(textid.getText().equals("") ||textpwd_before.getText().equals("") ||

textpwd_new1.getText().equals("")||textpwd_new2.getText().equals(""))
                            JOptionPane.showMessageDialog(null, "信息请输入完全");
else if (!textpwd_new1.getText().equals(textpwd_new2.getText()))
{
JOptionPane.showMessageDialog(null, "两次密码不同，请重新输入");
textpwd_new1.setText("");
textpwd_new2.setText("");
}
        else if (textpwd_new1.getText().equals(textpwd_new2.getText())&&
            textpwd_new1.getText().equals(textpwd_before.getText()))
                    {
                    JOptionPane.showMessageDialog(null, "新旧密码相同");
                                textpwd_new1.setText("");
                                textpwd_new2.setText("");
                        }
        else //信息正确进入数据库添加操作
        {
            try
```

```
{
//判断密码是否正确
String s_sql="SELECT Userpwd FROM UserInfo WHERE Userid='"
        +textid.getText()+"'";
System.out.println(s_sql);
ResultSet rs=con.stmt().executeQuery(s_sql);
rs.next();
if(!rs.getObject(1).equals(textpwd_before.getText()))
JOptionPane.showMessageDialog(null,"密码错误");
else
{
s_sql="UPDATE UserInfo SET Userpwd='"
    +textpwd_new1.getText()+"' WHERE Userid='"+textid.getText()+"'";
System.out.println(s_sql);
int count=con.stmt().executeUpdate(s_sql);
if (count==1)
JOptionPane.showMessageDialog(null,"修改成功");
else JOptionPane.showMessageDialog(null, "修改失败");
internalframe.dispose();
    }
} catch (Exception e1)
{e1.printStackTrace();}
    }
    }
        });
//取消按钮响应
button[1].addMouseListener(new MouseAdapter()
{
public void mouseClicked(MouseEvent e)
{
internalframe.dispose();
}
        });
internalframe.setVisible(true);
}
});
}
public void action_yhsc()//用户信息删除菜单响应
{
    menuItemyhsc.addActionListener(new ActionListener()
    {
    public void actionPerformed(ActionEvent e)
    {
        String[] s={"管理员","教师","学生"};
        JButton[] button={new JButton("查询"),new JButton("删除")};
        JLabel label1=new JLabel("用户名");
        JLabel label2=new JLabel("用户类型");
```

```
                final JTextField textid=new JTextField(10);
                final JComboBox jbox=new JComboBox(s);
    final JInternalFrame internalframe=new JInternalFrame("用户信息查询",true,true,
true,true);
                internalframe.setSize(310, 300);
                //加入 JTable
                Object[] column_names={"用户名","用户权限"};
                Object[][] data={};
            final DefaultTableModel model=new DefaultTableModel(data,column_names);
        final JTable tableview=new JTable();
        tableview.setModel(model);
        tableview.setPreferredScrollableViewportSize(new  Dimension(280,200));
    ableview.setAutoResizeMode(JTable.AUTO_RESIZE_SUBSEQUENT_COLUMNS);
                    JScrollPane spane=new JScrollPane(tableview);
                Container cc=internalframe.getContentPane();
                cc.setLayout(new FlowLayout(FlowLayout.RIGHT));
                cc.add(label1);
                cc.add(textid);
                cc.add(label2);
                cc.add(jbox);
                cc.add(button[0]);
                cc.add(button[1]);
                cc.add(spane);
                //查询响应事件
                button[0].addMouseListener(new MouseAdapter()
                    {
                public void mouseClicked(MouseEvent e)
                {
                try
                {
                ResultSet rs;
                //用户名域为空
                if (textid.getText().equals(""))
                {
            String s_sql="SELECT Userid,Userlevel FROM UserInfo WHERE Userlevel='"
                            +jbox.getSelectedItem()+"'";
                System.out.println(s_sql);
                rs=con.stmt().executeQuery(s_sql);
                rs.next();
                //返回值为空
                 if (rs.getRow()==0)
    JOptionPane.showMessageDialog(null,"没有权限为"+jbox.getSelectedItem()+"的
记录");
                //不为空，输出到 JTable 上
                else
                {
                //删除之前显示的行
```

```
            System.out.println(model.getRowCount());
            int count=model.getRowCount();
            for(int i=0;i<count;i++)
            model.removeRow(0);
            do
            {
            Vector vs=new Vector();
            for (int i=1;i<3;i++)
            vs.add(rs.getObject(i));
            model.addRow(vs);
            }while (rs.next());
            }
        }
        //用户名域不为空
        else{
        String s_sql="SELECT Userid,Userlevel FROM UserInfo WHERE Userid='"
            +textid.getText()+"'";
        System.out.println(s_sql);
        rs=con.stmt().executeQuery(s_sql);
        rs.next();
    //返回值为空
    if (rs.getRow()==0)
        JOptionPane.showMessageDialog(null, "没有用户名为"+textid.getText()+"
的记录");
    //不为空，输出到 JTable 上
    else
        {
    //删除之前显示的行
    System.out.println(model.getRowCount());
    int count=model.getRowCount();
    for(int i=0;i<count;i++)
    model.removeRow(0);
    do
    {
    Vector vs=new Vector();
    for (int i=1;i<3;i++)
    vs.add(rs.getObject(i));
    model.addRow(vs);
    }while (rs.next());
    }
    }
    } catch (Exception e1)
    {e1.printStackTrace();}}
    });
    //删除响应事件
    button[1].addMouseListener(new MouseAdapter()
    {
```

```
    public void mouseClicked(MouseEvent e)
    {
    int select=tableview.getSelectedRow();
     if (select==-1)
     JOptionPane.showMessageDialog(null, "请选择一条记录");
     else{//链接数据库删除记录
     int a=JOptionPane.showConfirmDialog(null, "确定删除? ");
    if (a==0)//确定删除，链接数据库删除选则条目
    {
try
                    {
                    String s_sql="DELETE FROM UserInfo"+
            WHERE Userid='"+tableview.getValueAt(select, 0)+"'";
                    System.out.println(s_sql);
                    int count=con.stmt().executeUpdate(s_sql);
                model.removeRow(select);//删除JTable表中的选择行
                    if (count==0)
                JOptionPane.showMessageDialog(null, "删除失败");
                else JOptionPane.showMessageDialog(null, "删除成功");
                        } catch (Exception e1)
                        {e1.printStackTrace();}
                        }
                    }

                }
            });
            Frmcjgl.desktop.add(internalframe);
            internalframe.setVisible(true);
        }
        }
        );
     }
     public void action_cxdl()//重新登录菜单响应
     {
        menuItemcxdl.addActionListener(new ActionListener()
        {
        public void actionPerformed(ActionEvent e)
        {
            JInternalFrame[]
allFrame=LoginApp.winmain.desktop.getAllFrames();
            for (int i=0;i<allFrame.length;i++)
            allFrame[i].dispose();
            Frmcjgl.app.dispose();
            LoginApp longin=new LoginApp();
        }
        });
     }
     public MenuItemxtgl()
```

```
        {
            //添加各个菜单项
            menuItemxtgl.add(menuItemyhgl);
            menuItemyhgl.add(menuItemyhcx);
            menuItemyhgl.add(menuItemyhtj);
            menuItemyhgl.add(menuItemyhxg);
            menuItemyhgl.add(menuItemyhsc);
            menuItemxtgl.add(menuItemcxdl);
            //响应各个菜单事件
            action_yhcx();
            action_yhtj();
            action_yhxg();
            action_yhsc();
            action_cxdl();
        }
        //类成员域
        JMenu menuItemxtgl=new JMenu("系统管理");
        JMenu menuItemyhgl=new JMenu("用户管理");
        JMenuItem menuItemyhcx=new JMenuItem("用户查询");
        JMenuItem menuItemyhtj=new JMenuItem("用户添加");
        JMenuItem menuItemyhsc=new JMenuItem("用户删除");
        JMenuItem menuItemyhxg=new JMenuItem("用户修改");
        JMenuItem menuItemcxdl=new JMenuItem("重新登录");
        final DataConnection con=new DataConnection();//链接数据库
    }
```

5. 学生信息管理类 MenuItemxsxxgl

MenuItemxsxxgl 类实现了学生管理菜单及其子菜单的功能，子菜单包括学生信息的查询功能（如图 10-9 所示）、添加功能（如图 10-10 所示）、修改功能（如图 10-11 所示）和删除功能（如图 10-12 所示）。

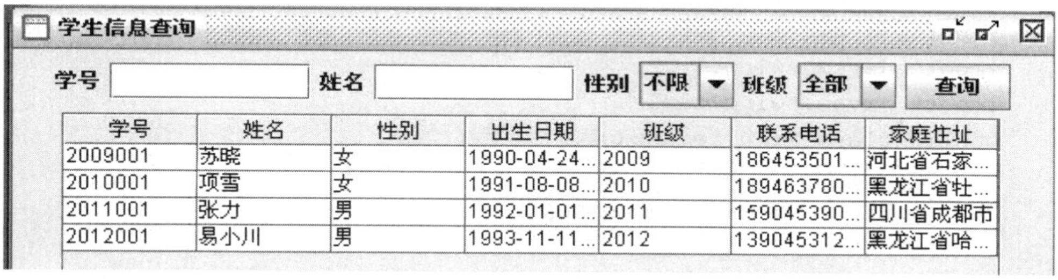
学生信息查询
学号
姓名
性别
不限
班级
全部
查询
学号　姓名　性别　出生日期　班级　联系电话　家庭住址
2009001　苏晓　女　1990-04-24...　2009　186453501...　河北省石家...
2010001　项雪　女　1991-08-08...　2010　189463780...　黑龙江省牡...
2011001　张力　男　1992-01-01...　2011　159045390...　四川省成都市
2012001　易小川　男　1993-11-11...　2012　139045312...　黑龙江省哈...

图 10-9　学生信息查询

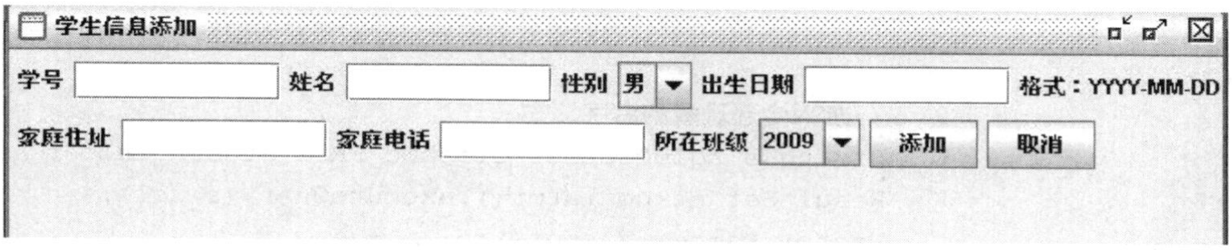
学生信息添加
学号
姓名
性别
男
出生日期
格式：YYYY-MM-DD
家庭住址
家庭电话
所在班级
2009
添加
取消

图 10-10　学生信息添加

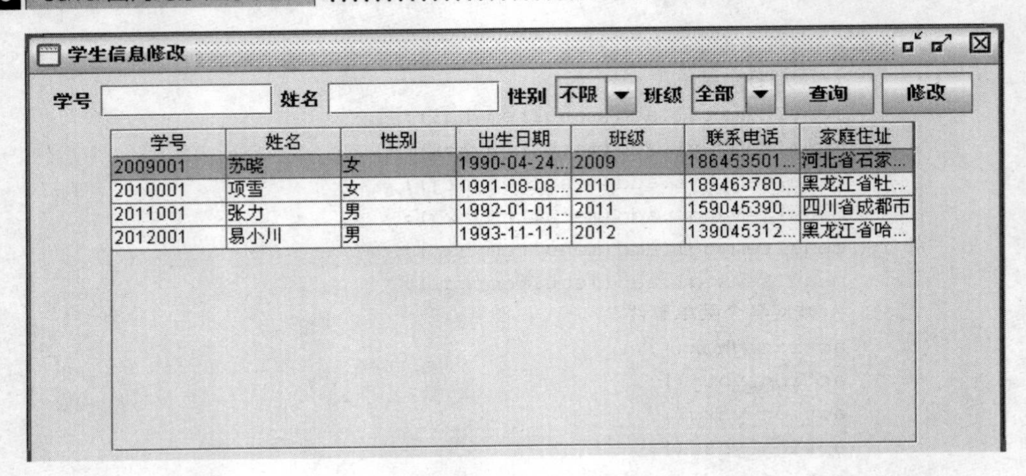

图 10-11　学生信息修改

图 10-12　学生信息删除

实现代码如下：

```
import java.awt.*;
import java.awt.event.*;
import java.sql.*;
import java.util.Vector;
import javax.swing.*;
import javax.swing.table.*;
public class MenuItemxsxxgl {
    //学生信息查询
    public void menuitemxsxxcx()
    {
        menuItemxsxxcx.addActionListener(new ActionListener()
        {
        public void actionPerformed(ActionEvent e)
        {
            try{
                //加入全部班级的名字
                String s_sql="SELECT Classid FROM StudentInfo";
                ResultSet st=con.stmt().executeQuery(s_sql);
                Vector vs=new Vector();
                vs.add("全部");
```

```
                    while(st.next())
                    {
                        vs.add(st.getString(1));
                    }
                    String[] s={"不限","男","女"};
    Object[] column_names={"学号","姓名","性别","出生日期","班级","联系电话","家
庭住址" };
                    Object[][] data={};
        final                    DefaultTableModel                         model=new
DefaultTableModel(data,column_names);
                    final JTable tableview=new JTable();
                    tableview.setModel(model);
        tableview.setPreferredScrollableViewportSize(new
Dimension(530,200));

    tableview.setAutoResizeMode(JTable.AUTO_RESIZE_SUBSEQUENT_COLUMNS);
                    JScrollPane spane=new JScrollPane(tableview);
    final JInternalFrame internalframe=
    new JInternalFrame("学生信息查询",true,true,true,true);
                    internalframe.setSize(600, 300);
                    Frmcjgl.desktop.add(internalframe);
    JLabel[] label={new JLabel("学号"),new JLabel("姓名"),new JLabel("性别"),
new JLabel("班级")};
                JButton button=new JButton("查询");
                final JTextField textsid=new JTextField(10);
                final JTextField textsname=new JTextField(10);
                final JComboBox boxsex=new JComboBox(s);
                final JComboBox boxclassid=new JComboBox(vs);
                Container cc=internalframe.getContentPane();
                cc.setLayout(new FlowLayout(FlowLayout.CENTER));
                cc.add(label[0]);
                cc.add(textsid);
                cc.add(label[1]);
                cc.add(textsname);
                cc.add(label[2]);
                cc.add(boxsex);
                cc.add(label[3]);
                cc.add(boxclassid);
                cc.add(button);
                cc.add(spane);
                //学号文本框只能输入数字
                textsid.addKeyListener(new KeyAdapter()
                {
                    public void keyTyped(KeyEvent e)
                    {
                    if ((e.getKeyChar()<'0')||(e.getKeyChar()>'9'))
                        e.consume();
```

```
            }
        });
            //查询按钮
        button.addMouseListener(new MouseAdapter()
        {
            public void mouseClicked(MouseEvent e)
            {
            try{
                //删除之前显示的行
                int count=model.getRowCount();
                for(int i=0;i<count;i++)
                model.removeRow(0);
                //构造 sql 查询语句
            String s_sid,s_sname,s_sex,s_classid;
            if (textsid.getText().equals("")) s_sid="";
            else s_sid=" Sno='"+textsid.getText()+"' and";
            if (textsname.getText().equals("")) s_sname="";
            else s_sname=" Sname='"+textsname.getText()+"' and";
         if (boxsex.getSelectedItem().equals("不限")) s_sex="";
            else s_sex=" Sex='"+boxsex.getSelectedItem()+"' and";
        if (boxclassid.getSelectedItem().equals("全部")) s_classid="";
         else
         s_classid=" Classid='"+boxclassid.getSelectedItem()+"'";
 String s_sql="SELECT * FROM StudentInfo WHERE "+s_sid+s_sname+s_sex+s_classid;
        StringBuffer s=new StringBuffer(s_sql);
        if (s.charAt(s.length()-1)=='d')
        {
        s.setCharAt(s.length()-1,' ');
        s.setCharAt(s.length()-2,' ');
        s.setCharAt(s.length()-3,' ');
        }
        s_sql=s.toString();
       if (textsid.getText().equals("")&& textsname.getText().equals("")
  &&boxsex.getSelectedItem().equals("不限") && boxclassid.getSelectedItem().
equals("全部"))
        s_sql="SELECT * FROM StudentInfo";
        System.out.println(s_sql);
        ResultSet rs=con.stmt().executeQuery(s_sql);
         while(rs.next())
         {
         Vector vs=new Vector();
         for(int i=0;i<7;i++)
         vs.add(rs.getObject(i+1));
         model.addRow(vs);
      }
        } catch (Exception e1)
        {e1.printStackTrace();}
```

```
                }
            });
            internalframe.setVisible(true);
        } catch (Exception e1)
        {e1.printStackTrace();}
        }
        });
    }
    //学生信息添加
    public void menuitemxsxxtj()
    {
        menuItemxsxxtj.addActionListener(new ActionListener()
        {
        public void actionPerformed(ActionEvent e)
        {
            try{
            //加入全部班级的名字
            String s_sql="SELECT Classid FROM StudentInfo";
            ResultSet st=con.stmt().executeQuery(s_sql);
            Vector vs=new Vector();
            while(st.next())
            {
                vs.add(st.getString(1));
            }
            String[] s={"男","女"};
  JLabel[] label={new JLabel("学号"),new JLabel("姓名"),new JLabel("性别"),
new JLabel("出生日期"),new JLabel("格式: YYYY-MM-DD"),new JLabel("家庭住址"),
new JLabel("家庭电话"),new JLabel("所在班级")};
            final JTextField textsid=new JTextField(10);
            final JTextField textsname=new JTextField(10);
            final JTextField textbirthday=new JTextField(10);
            final JTextField textaddress=new JTextField(10);
            final JTextField texttel=new JTextField(10);
            final JComboBox boxsex=new JComboBox(s);
            final JComboBox boxclassid=new JComboBox(vs);
            JButton[] button={new JButton("添加"),new JButton("取消")};
              final JInternalFrame internalframe=
  new JInternalFrame("学生信息添加",true,true,true,true);
            internalframe.setSize(680, 200);
            Frmcjgl.desktop.add(internalframe);
            Container cc=internalframe.getContentPane();
            cc.setLayout(new FlowLayout(FlowLayout.LEFT));
            cc.add(label[0]);
            cc.add(textsid);
            cc.add(label[1]);
            cc.add(textsname);
            cc.add(label[2]);
```

```
            cc.add(boxsex);
            cc.add(label[3]);
            cc.add(textbirthday);
            cc.add(label[4]);
            cc.add(label[5]);
            cc.add(textaddress);
            cc.add(label[6]);
            cc.add(texttel);
            cc.add(label[7]);
            cc.add(boxclassid);
            cc.add(button[0]);
            cc.add(button[1]);
            //学号文本框只能输入数字
            textsid.addKeyListener(new KeyAdapter()
            {
                public void keyTyped(KeyEvent e)
                {
                    if ((e.getKeyChar()<'0')||(e.getKeyChar()>'9'))
                            e.consume();
                }
            });
            //联系电话文本框只能输入数字
            texttel.addKeyListener(new KeyAdapter()
            {
                public void keyTyped(KeyEvent e)
                {
                    if ((e.getKeyChar()<'0')||(e.getKeyChar()>'9'))
                            e.consume();
                }
            });
            //按下添加按钮时
            button[0].addMouseListener(new MouseAdapter()
            {
                public void mouseClicked(MouseEvent e)
                {
String s_sql="SELECT * FROM StudentInfo WHERE Sno='"+textsid.getText()+"'";
                    System.out.println(s_sql);
                    ResultSet rs;
                    int flag=0;
                        try {
                            rs = con.stmt().executeQuery(s_sql);
                            rs.next();
                            flag=rs.getRow();
                        } catch (SQLException e2)
      {e2.printStackTrace();}
     if (flag!=0) JOptionPane.showMessageDialog(null, "已有记录，无法插入");
                    //插入
```

```
            else if (!textsid.getText().equals("") && !textsname.getText().equals("")

    && !textbirthday.getText().equals("")&& !textaddress.getText().equals("")
            && !texttel.getText().equals(""))
            {
             try {
    s_sql="INSERT INTO StudentInfo(Sno,Sname,Sex,Birthday,Classid,Tel,Address)" +
    "
VALUES('"+textsid.getText()+"','"+textsname.getText()+"','"+boxsex.getSelectedItem()

    +"','"+textbirthday.getText()+"','"+boxclassid.getSelectedItem()+"','"
             +texttel.getText()+"','"+textaddress.getText()+"')";
            System.out.println(s_sql);
             int count=con.stmt().executeUpdate(s_sql);
             if (count!=0)
             JOptionPane.showMessageDialog(null, "插入成功");
             else JOptionPane.showMessageDialog(null, "插入失败");
             } catch (SQLException e1)
             {e1.printStackTrace();}
            }
            else  JOptionPane.showMessageDialog(null, "请输入完全");
            }
            });
          //按下取消按钮时
          button[1].addMouseListener(new MouseAdapter()
            {
            public void mouseClicked(MouseEvent e)
             {
             internalframe.dispose();
             }
            });
            internalframe.setVisible(true);
             } catch (SQLException e2)
                   {e2.printStackTrace();}
             }
            });
        }
        //学生信息修改
        public void menuitemxsxxxg()
        {
            menuItemxsxxxg.addActionListener(new ActionListener()
            {
            public void actionPerformed(ActionEvent e)
             {
                 try{
                      //加入全部班级的名字
                      String s_sql="SELECT Classid FROM StudentInfo";
```

```
                    ResultSet st=con.stmt().executeQuery(s_sql);
                    Vector vs=new Vector();
                    vs.add("全部");
                    while(st.next())
                    {
                        vs.add(st.getString(1));
                    }
                    String[] s={"不限","男","女"};
                    //加入 JTable
    Object[] column_names={"学号","姓名","性别","出生日期","班级","联系电话","家
庭住址" };
                    Object[][] data={};
    final DefaultTableModel model=new DefaultTableModel(data,column_names);
                final JTable tableview=new JTable();
                tableview.setModel(model);
    tableview.setPreferredScrollableViewportSize(new Dimension(530,200));
        tableview.setAutoResizeMode(JTable.AUTO_RESIZE_SUBSEQUENT_COLUMNS);
        JScrollPane spane=new JScrollPane(tableview);
    final JInternalFrame internalframe=new JInternalFrame("学生信息修改",true,
true,true,true);
                    internalframe.setSize(650, 300);
                    Frmcjgl.desktop.add(internalframe);
    JLabel[] label={new JLabel("学号"),new JLabel("姓名"),new JLabel("性别
"),new JLabel("班级")};
                    JButton[] button={new JButton("查询"),new JButton("修改")};
                    final JTextField textsid=new JTextField(10);
                    final JTextField textsname=new JTextField(10);
                    final JComboBox boxsex=new JComboBox(s);
                    final JComboBox boxclassid=new JComboBox(vs);
                    Container cc=internalframe.getContentPane();
                    cc.setLayout(new FlowLayout(FlowLayout.CENTER));
                    cc.add(label[0]);
                    cc.add(textsid);
                    cc.add(label[1]);
                    cc.add(textsname);
                    cc.add(label[2]);
                    cc.add(boxsex);
                    cc.add(label[3]);
                    cc.add(boxclassid);
                    cc.add(button[0]);
                    cc.add(button[1]);
                    cc.add(spane);
                    //学号文本框只能输入数字
                    textsid.addKeyListener(new KeyAdapter()
                    {
                        public void keyTyped(KeyEvent e)
                        {
```

```
                if ((e.getKeyChar()<'0')||(e.getKeyChar()>'9'))
                        e.consume();
                        }
                    });
                    //查询按钮
                button[0].addMouseListener(new MouseAdapter()
                {
                    public void mouseClicked(MouseEvent e)
                    {
                        try{
                        //删除之前显示的行
                        int count=model.getRowCount();
                        for(int i=0;i<count;i++)
                        model.removeRow(0);
                        //构造 sql 查询语句
                String s_sid,s_sname,s_sex,s_classid;
                if (textsid.getText().equals("")) s_sid="";
                else s_sid=" Sno='"+textsid.getText()+"' and";
                if (textsname.getText().equals("")) s_sname="";
                else s_sname=" Sname='"+textsname.getText()+"' and";
                if (boxsex.getSelectedItem().equals("不限")) s_sex="";
                else s_sex=" Sex='"+boxsex.getSelectedItem()+"' and";
            if (boxclassid.getSelectedItem().equals("全部")) s_classid="";
                else
    s_classid=" Classid='"+boxclassid.getSelectedItem()+"'";
  String s_sql="SELECT * FROM StudentInfo WHERE "+s_sid+s_sname+s_sex+s_classid;
                StringBuffer s=new StringBuffer(s_sql);
                if (s.charAt(s.length()-1)=='d')
                    {
                s.setCharAt(s.length()-1,' ');
                s.setCharAt(s.length()-2,' ');
                s.setCharAt(s.length()-3,' ');
                    }
                s_sql=s.toString();
  if(textsid.getText().equals("")&&textsname.getText().equals("")
  && boxsex.getSelectedItem().equals("不限") && boxclassid.getSelectedItem().
equals("全部"))
      s_sql="SELECT * FROM StudentInfo";
                System.out.println(s_sql);
                ResultSet rs=con.stmt().executeQuery(s_sql);
                    while(rs.next())
                {
                Vector vs=new Vector();
                for(int i=0;i<7;i++)
                vs.add(rs.getObject(i+1));
                model.addRow(vs);
                }
```

```
      } catch (Exception e1)
      {
     e1.printStackTrace();
      }
      }
     });
    //修改按钮
         button[1].addMouseListener(new MouseAdapter()
         {
          public void mouseClicked(MouseEvent e)
      {
      int select_rowcount=tableview.getSelectedRowCount();
      final int select_row=tableview.getSelectedRow();
      if (select_rowcount!=1)
      JOptionPane.showMessageDialog(null, "请选择一行");
      else
      {
       try{
      //加入全部班级的名字
      String s_sql="SELECT Classid FROM StudentInfo";
      ResultSet st=con.stmt().executeQuery(s_sql);
      Vector vs=new Vector();
         while(st.next())
          {
          vs.add(st.getString(1));
         }
          String[] s={"男","女"};
JLabel[] label={new JLabel("学号"),new JLabel("姓名"),new JLabel("性别"),
new JLabel("出生日期"),new JLabel("格式: YYYY-MM-DD"),new JLabel("家庭住址"),
new JLabel("家庭电话"),new JLabel("所在班级")};
         final JTextField textsid=new JTextField(10);
         final JTextField textsname=new JTextField(10);
         final JTextField textbirthday=new JTextField(10);
         final JTextField textaddress=new JTextField(10);
         final JTextField texttel=new JTextField(10);
         final JComboBox boxsex=new JComboBox(s);
         final JComboBox boxclassid=new JComboBox(vs);
         JButton[] button={new JButton("保存"),new JButton("取消")};
final JInternalFrame internalframe=
new JInternalFrame("学生信息保存",true,true,true,true);
         internalframe.setSize(680, 200);
         Frmcjgl.desktop.add(internalframe);
         Container cc=internalframe.getContentPane();
         cc.setLayout(new FlowLayout(FlowLayout.LEFT));
         cc.add(label[0]);
         cc.add(textsid);
      cc.add(label[1]);
```

```
cc.add(textsname);
cc.add(label[2]);
cc.add(boxsex);
cc.add(label[3]);
cc.add(textbirthday);
cc.add(label[4]);
cc.add(label[5]);
cc.add(textaddress);
cc.add(label[6]);
cc.add(texttel);
cc.add(label[7]);
cc.add(boxclassid);
cc.add(button[0]);
cc.add(button[1]);
//学号文本框只能输入数字
textsid.addKeyListener(new KeyAdapter()
{
public void keyTyped(KeyEvent e)
{
f ((e.getKeyChar()<'0')||(e.getKeyChar()>'9'))
e.consume();
}
});
textsid.setEditable(false);
textsid.setText((String)tableview.getValueAt(select_row, 0));
//联系电话文本框只能输入数字
texttel.addKeyListener(new KeyAdapter()
{
public void keyTyped(KeyEvent e)
{
if ((e.getKeyChar()<'0')||(e.getKeyChar()>'9'))
e.consume();
}
});
//按下保存按钮时
button[0].addMouseListener(new MouseAdapter()
{
public void mouseClicked(MouseEvent e)
{
try {
    String s_sql="UPDATE StudentInfo SET Sname='"+textsname.getText()
+"',Sex='"+boxsex.getSelectedItem()+"',Birthday='"+textbirthday.getText
()+"',Classid='"+boxclassid.getSelectedItem()+"',Tel='"+texttel.getText()+
"',Address='"+textaddress.getText()+"' WHERE Sno='"+textsid.getText()+"'";
    System.out.println(s_sql);
    int count=con.stmt().executeUpdate(s_sql);
    if (count!=0){
```

```
JOptionPane.showMessageDialog(null, "添加成功");
model.setValueAt(textsname.getText(), select_row, 1);
model.setValueAt(boxsex.getSelectedItem(), select_row, 2);
model.setValueAt(textbirthday.getText(), select_row, 3);
model.setValueAt(boxclassid.getSelectedItem(), select_row, 4);
model.setValueAt(texttel.getText(), select_row, 5);
model.setValueAt(textaddress.getText(), select_row, 6);
internalframe.dispose();
            }
else JOptionPane.showMessageDialog(null, "添加失败");
} catch (SQLException e1) {
e1.printStackTrace();}
           }
         });
      //按下取消按钮时
      button[1].addMouseListener(new MouseAdapter()
      {
          public void mouseClicked(MouseEvent e)
         {
         internalframe.dispose();
      }
      });
      internalframe.setVisible(true);
      } catch (SQLException e2)
      {e2.printStackTrace();}
      }}
      });
      internalframe.setVisible(true);
      } catch (Exception e1)
         {e1.printStackTrace();}
      }
       });
      }
//学生信息删除
public void menuitemxsxxsc()
{
    menuItemxsxxsc.addActionListener(new ActionListener()
    {
    public void actionPerformed(ActionEvent e)
    {
        try{
            //加入全部班级的名字
            String s_sql="SELECT Classid FROM StudentInfo";
            ResultSet st=con.stmt().executeQuery(s_sql);
            Vector vs=new Vector();
            vs.add("全部");
            while(st.next())
```

```
                    {
                        vs.add(st.getString(1));
                    }
                    String[] s={"不限","男","女"};

                    //加入 JTable
    Object[] column_names={"学号","姓名","性别","出生日期","班级","联系电话","家
庭住址" };
                Object[][] data={};
        final DefaultTableModel model=neDefaultTableModel(data,column_names);
                    final JTable tableview=new JTable();
                    tableview.setModel(model);
        tableview.setPreferredScrollableViewportSize(new Dimension(530,200));

    tableview.setAutoResizeMode(JTable.AUTO_RESIZE_SUBSEQUENT_COLUMNS);
                    JScrollPane spane=new JScrollPane(tableview);
                final JInternalFrame internalframe=
    new JInternalFrame("学生信息删除",true,true,true,true);
                internalframe.setSize(650, 300);
                Frmcjgl.desktop.add(internalframe);
    JLabel[] label={new JLabel("学号"),new JLabel("姓名"),new JLabel("性别
"),new JLabel("班级")};
                Button[] button={new JButton("查询"),new JButton("删除")};
                final JTextField textsid=new JTextField(10);
                final JTextField textsname=new JTextField(10);
                final JComboBox boxsex=new JComboBox(s);
                final JComboBox boxclassid=new JComboBox(vs);
                Container cc=internalframe.getContentPane();
                cc.setLayout(new FlowLayout(FlowLayout.CENTER));
                cc.add(label[0]);
                cc.add(textsid);
                cc.add(label[1]);
                cc.add(textsname);
                cc.add(label[2]);
                cc.add(boxsex);
                cc.add(label[3]);
                cc.add(boxclassid);
                cc.add(button[0]);
                cc.add(button[1]);
                cc.add(spane);
                //学号文本框只能输入数字
                textsid.addKeyListener(new KeyAdapter()
                {
                public void keyTyped(KeyEvent e)
                {
                if ((e.getKeyChar()<'0')||(e.getKeyChar()>'9'))
                    e.consume();
```

```
        }
        });
        //查询按钮
        button[0].addMouseListener(new MouseAdapter()
        {
        public void mouseClicked(MouseEvent e)
        {
        try{
        //删除之前显示的行
        int count=model.getRowCount();
        for(int i=0;i<count;i++)
        model.removeRow(0);
        //构造sql查询语句
        String s_sid,s_sname,s_sex,s_classid;
        if (textsid.getText().equals("")) s_sid="";
        else s_sid=" Sno='"+textsid.getText()+"' and";
        if (textsname.getText().equals("")) s_sname="";
        else s_sname=" Sname='"+textsname.getText()+"' and";
        if (boxsex.getSelectedItem().equals("不限")) s_sex="";
        else s_sex=" Sex='"+boxsex.getSelectedItem()+"' and";
        if (boxclassid.getSelectedItem().equals("全部")) s_classid="";
        else
    s_classid=" Classid='"+boxclassid.getSelectedItem()+"'";
         String s_sql="SELECT * FROM StudentInfo WHERE "+s_sid+s_sname+s_
         sex+s_classid;
        StringBuffer s=new StringBuffer(s_sql);
        if (s.charAt(s.length()-1)=='d')
        {
        s.setCharAt(s.length()-1,' ');
        s.setCharAt(s.length()-2,' ');
        s.setCharAt(s.length()-3,' ');
        }
        s_sql=s.toString();

 if(textsid.getText().equals("")&&textsname.getText().equals("")&&
 boxsex.getSelectedItem().equals("不限") &&
 boxclassid.getSelectedItem().equals("全部"))
        s_sql="SELECT * FROM StudentInfo";
        System.out.println(s_sql);
        ResultSet rs=con.stmt().executeQuery(s_sql);
           while(rs.next())
        {
        Vector vs=new Vector();
        for(int i=0;i<7;i++)
        vs.add(rs.getObject(i+1));
        model.addRow(vs);
        }
```

```
        } catch (Exception e1)
        {e1.printStackTrace();}
        }
            });
        //删除按钮
        button[1].addMouseListener(new MouseAdapter()
        {
        public void mouseClicked(MouseEvent e)
        {
        int select_rowcount=tableview.getSelectedRowCount();
        final int select_row=tableview.getSelectedRow();
        if (select_rowcount!=1)
        JOptionPane.showMessageDialog(null, "请选择一行");
        else
        {
        int a=JOptionPane.showConfirmDialog(null, "确定删除? ");
        if (a==0)//确定删除，链接数据库删除选则条目
        {
        try
           {
        String s_sql="DELETE FROM StudentInfo"+
        " WHERE Sno='"+tableview.getValueAt(select_row, 0)+"'";
        System.out.println(s_sql);
        int count=con.stmt().executeUpdate(s_sql);
        if (count==0)
        JOptionPane.showMessageDialog(null, "删除失败");
        else
        {
        JOptionPane.showMessageDialog(null, "删除成功");
          model.removeRow(select_row);//删除 JTable 表中的选择行
        }
           } catch (Exception e1)
          e1.printStackTrace();}
           }
       }
    }
    });
   internalframe.setVisible(true);
        } catch (Exception e1)
            {e1.printStackTrace();}
   }
  });
}
        public MenuItemxsxxgl()
            {
            menuItemxsxxgl.add(menuItemxsxxcx);
            menuItemxsxxgl.add(menuItemxsxxtj);
```

```
        menuItemxsxxgl.add(menuItemxsxxxg);
                menuItemxsxxgl.add(menuItemxsxxsc);
                //响应事件
                menuitemxsxxcx();
                menuitemxsxxtj();
                menuitemxsxxxg();
                menuitemxsxxsc();
            }
                JMenu menuItemxsxxgl=new JMenu("学生管理");
                JMenuItem menuItemxsxxcx=new JMenuItem("学生信息查询");
                JMenuItem menuItemxsxxtj=new JMenuItem("学生信息添加");
                JMenuItem menuItemxsxxxg=new JMenuItem("学生信息修改");
                JMenuItem menuItemxsxxsc=new JMenuItem("学生信息删除");
                final DataConnection con=new DataConnection();//链接数据库
}
```

6．系统退出类 MenuItemtc

MenuItemtc 类实现了退出本系统的功能。

实现代码如下：

```
import java.awt.event.*;
import javax.swing.*;
public class MenuItemtc {//退出菜单类
    static JMenu menuItemtc=new JMenu("退出");
    static JMenuItem tc=new JMenuItem("退出程序");
    public MenuItemtc()
      {
        menuItemtc.add(tc);
        tc.addActionListener(new ActionListener()
            {
                public void actionPerformed(ActionEvent e)
                {
                    System.exit(0);
                }
            }
        );
      }
}
```

参 考 文 献

[1] 高飞，陆佳炜，徐俊. Java 程序设计实用教程［M］. 北京：清华大学出版社，2013.

[2] 耿祥义，张跃平. Java 程序设计实用教程［M］. 北京：人民邮电出版社，2015.

[3] 霍斯特曼，科内尔. Java 核心技术，卷 1：基础知识［M］. 北京：机械工业出版社，2014.

[4] Bruce Eckel. Java 编程思想［M］. 陈昊鹏译. 北京：机械工业出版社，2007.

[5] 传智播客高教产品研发部. Java 基础入门［M］. 北京：清华大学出版社，2014.

[6] 张席. JAVA 语言程序设计教程［M］. 西安：西安电子科技大学出版社，2015.